“十三五”高等职业教育规划教材

计算机专业英语

（第三版）

张洪颖　主编

董晓霞　王成霞　袁俊娥　孙　丹　副主编

中国铁道出版社有限公司
CHINA RAILWAY PUBLISHING HOUSE CO., LTD.

内 容 简 介

本书（英/汉对照）内容涉及计算机硬件、计算机软件、程序设计语言、数据库、多媒体、计算机网络等方面的知识以及科技英语的翻译技巧和语法知识。为了方便教学，每章都列出了专业词汇对照表、语言点注释及练习题；正文后还给出了每一章的练习题参考答案和参考译文。本书的内容覆盖面较广，建议教师在教学过程中根据教学的实际需要和学生的具体情况适当进行内容取舍。

本书适合作为高职高专院校计算机专业及相关专业的计算机专业英语课程教材，也可供计算机相关专业技术人员及其他有兴趣的读者学习和参考。

图书在版编目（CIP）数据

计算机专业英语/张洪颖主编. —3版. —北京：中国铁道出版社，2018.6(2019.12重印)
“十三五”高等职业教育规划教材
ISBN 978-7-113-24587-0

Ⅰ.①计… Ⅱ.①张… Ⅲ.①电子计算机-英语-高等职业教育-教材 Ⅳ.①TP3

中国版本图书馆CIP数据核字(2018)第123262号

书　　名：计算机专业英语（第三版）
作　　者：张洪颖　主编

策　　划：翟玉峰　　　　**读者热线：**（010）63550836
责任编辑：翟玉峰　田银香
封面设计：付　巍
封面制作：刘　颖
责任校对：张玉华
责任印制：郭向伟

出版发行：中国铁道出版社有限公司（100054，北京市西城区右安门西街8号）
网　　址：http://www.tdpress.com/51eds/
印　　刷：三河市宏盛印务有限公司
版　　次：2009年7月第1版　2014年4月第2版　2018年6月第3版　2019年12月第2次印刷
开　　本：787 mm×1 092 mm　1/16　**印张：**12　**字数：**282千
印　　数：2 001～3 500册
书　　号：ISBN 978-7-113-24587-0
定　　价：34.00元

PREFACE

第三版前言>>>

我国高等职业教育发展很快，《国家中长期教育改革和发展规划纲要（2010—2020）》明确提出要大力发展职业教育，国家和政府高度重视职业教育的发展。高职英语课程是高等职业教育各专业的一门必修课。目前，随着高等职业教育教学改革的推进，高职公共英语正在实现向职业英语的转变。很多学校都开设了专门用途英语，有的已经用专门用途英语代替了基础英语。在此情况下，选择一本适合高职学生特点的专门用途英语教材就显得特别重要。

《计算机专业英语》自2009年首次印刷出版以来，得到了专家和广大读者的肯定与厚爱，2010年被**教育部高等学校高职高专计算机类专业教学指导委员会评为优秀教材**。第一版累计印刷量达8 000册，第二版累计印刷量达7 500册。本书本着因材施教的教学理念编写，坚持“实用为主，够用为度”的方针。经过市场的检验，本书难度适合高职学生，教材编写模式符合高职学生特点，适于高职学生使用。

本书第三版基本框架和理念不变，依然坚持“实用为主，够用为度”的方针，以语言基本技能训练和培养实际从事涉外专业活动和工作需求的语言应用能力为出发点，结合学生专业知识，力求提高学生未来工作岗位所需要的专业英语技能；依然保持了前两版原有特色——教材编排采用任务驱动的模式，以任务为线索，每章通过一篇对话提出一个实际任务，然后引入正文——相关专业知识，最后在练习里回应开篇对话提出的任务——给出解决问题的方案。

本次修订是在第二版的基础上更新了陈旧的内容，重点更新了第二章计算机软件的内容。本书由北京联合大学张洪颖任主编，董晓霞、王成霞、袁俊娥和长春广播电视大学孙丹任副主编。

本书的主要读者对象是计算机专业及相关专业的高职高专学生和从事计算机相关工作的专业人员及其他有兴趣的读者。

鉴于编者水平有限，疏漏和不妥之处在所难免，恳请广大读者批评指正。

编 者

2018年2月

第一版前言

计算机行业是国际化的行业。目前，该行业国际化程度日益提高，国际 IT 企业在国内的研发基地的建设、欧美软件外包的迅猛发展以及中国 IT 企业国际化进程的加快，都对国际化 IT 人才的培养提出了迫切要求，软件外包业的发展尤其明显。软件制造业、高科技研发逐渐向中国转移，然而外语语言问题却成为其发展瓶颈，缺乏既掌握专业技术又精通英语的人才。

针对目前的状况，高职院校要想培养出 IT 企业真正需要的人才，就必须加强学生在计算机专业英语方面的学习。本教材本着“以实用为主，突出应用”的高职人才培养的总体目标，坚持“实用为主，够用为度”的方针，以语言基本技能训练和培养实际从事涉外专业活动和工作需求的语言应用能力为出发点，结合学生专业知识，力求提高学生未来工作岗位所需要的专业英语技能。

本教材的编排渗透了任务驱动的模式。以任务为线索，每章通过一篇对话提出一个实际任务，然后引入正文——相关专业知识，最后在练习里回应开篇对话提出的任务——给出解决方案，解决问题。

本书每章除了正文外，还列出了专业词汇对照表及语言点注释，正文后还附有练习题。另外，教材后面还有每一章的练习题参考答案和参考译文，可供教师教学时参考，也可供自学者使用。书的内容覆盖面较广，建议教师在教学过程中根据教学的实际要求和学生的具体情况适当进行取舍。

本书由张洪颖主编，董晓霞、王成霞、袁俊娥副主编，张淑艳参与了编写。

本书的主要读者对象是计算机专业及相关专业的高职、高专学生和从事计算机相关工作的专业人员及其他有兴趣的读者。

由于编者水平有限，加之时间仓促，书中难免有不足与疏漏之处，欢迎广大读者批评指正。

编 者

2009 年 2 月

CONTENTS（目录）

Chapter 1 Computer Hardware

Reading

Dialogue

What can I do?

Background: *Richard and Jim were classmates in high school and they are also very good friends. Now Richard is majoring in computer science in a university while Jim majors in business trade. Jim wants to buy a computer, but he knows little about it, so he wants Richard to introduce him something about computer hardware. They talked about this on the phone before and made this appointment.*

Richard: Hi. Haven't seen you for a long time. How is everything going?

Jim: So far so good. And you?

Richard: I'm fine.

Jim: Thank you for lending me a hand.

Richard: You are welcome. Have you decided whether you would buy a brand machine or DIY (Do It Yourself)?

Jim: I don't know for sure.

Richard: Do you have any idea of CPU, input devices, output devices and storage?

Jim: You know, I'm a layman. I am at a loss when hearing these specialized words and phrases.

Richard: In my opinion, first of all, you should have some knowledge of computer hardware, especially if you want to DIY. I've brought you some materials. You'd better have a look. Then we'll make a further decision.

Jim: OK. That's just what I think. After going through the materials, I'll call you or we may meet somewhere.

Richard: That's all right. See you.

Jim: See you.

Here is the material:

A computer is a fast and accurate system that is organized to accept, store and process data, and produce results under the direction of a stored program.

A PC system consists of two basic parts — hardware and software. Hardware is the physical part of the system that can be seen and touched while software refers to programs that control the operation of the hardware.

Computer hardware can be divided into four categories: CPU, storage devices, input devices and output devices.

Fig. 1-1-1 shows the basic organization of a computer system.

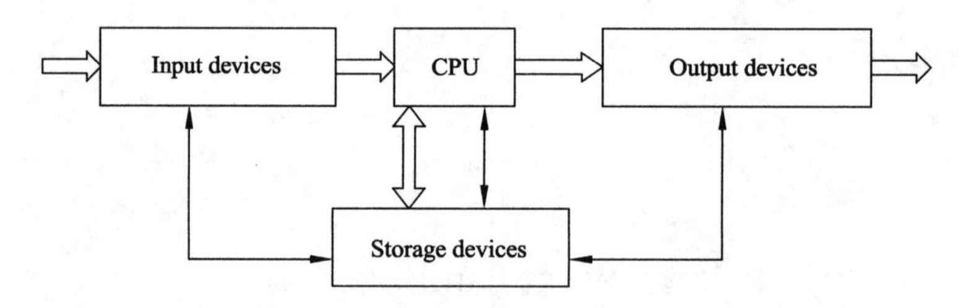

Fig. 1-1-1 A computer system

1.1 Input and Output Devices

How do you put data to the CPU? How do you get information out? Here we will introduce the devices which enable people and computer to communicate. Input devices translate data and program instructions that people understand into a form that computers can process. Output devices do the opposite. They translate computer-processed information into a form that people can comprehend.

Input Devices

There are several ways to get new information or input into a computer. To serve different application purposes，input devices can be classified into four types: letter input devices, pointing input devices, imaging and video input devices, and audio input devices.

The typical letter input device is keyboard. It is also one of the most common input devices. The keyboard has keys for characters (letters, numbers and punctuation marks) and special commands. Pressing the keys tells the computer what to do or what to write. Generally, the traditional 101-key keyboard has four key groups: the function key row at the top of the keyboard, the typewriter keypad, the cursor-edit keypad with arrows indicating up, down, left and right, and the numeric keypad (Fig.1-1-2). The new 104-key keyboard adds three shortcut keys for Windows (Fig.1-1-3). Besides the traditional design, there are also folding keyboard (Fig.1-1-4) and ergonomic keyboard (Fig.1-1-5). Mostly used brands are Logitech, Microsoft, Philips, ViewSonic, Shuangfeiyan, etc.

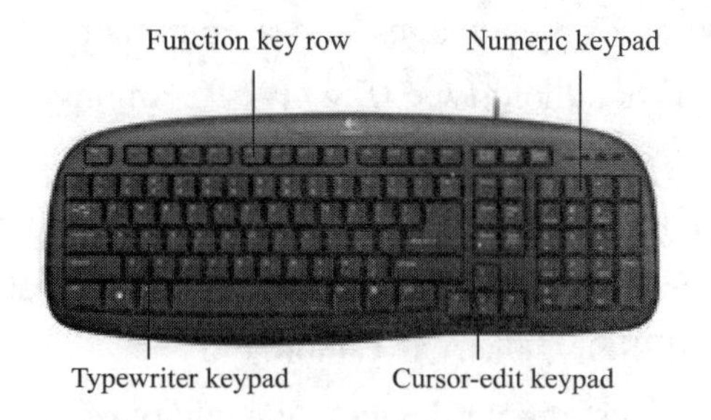

Fig. 1-1-2 101-key keyboard

Fig. 1-1-3 104-key keyboard

Fig. 1-1-4 Folding keyboard

Fig. 1-1-5 Ergonomic keyboard

The most widely used pointing device is the mouse. A mouse controls a pointer on the monitor which usually appears in the shape of an arrow. A mouse can have one, two or three buttons, which are used to select command options and to control information presented on the monitor. There are three basic types of mouse: first, mechanical mouse (Fig. 1-1-6). It has a ball at the bottom, and rolling it on a smooth surface can control the pointer on the screen. Second, optical mouse (Fig. 1-1-7), which is the mostly widely used now. It emits and senses light to detect mouse movement. It can be used on any surface and is more precise. Third, cordless or wireless mouse (Fig. 1-1-8). It uses radio waves or infrared light waves to communicate with the system unit. Mostly used brands are similar to those of keyboards.

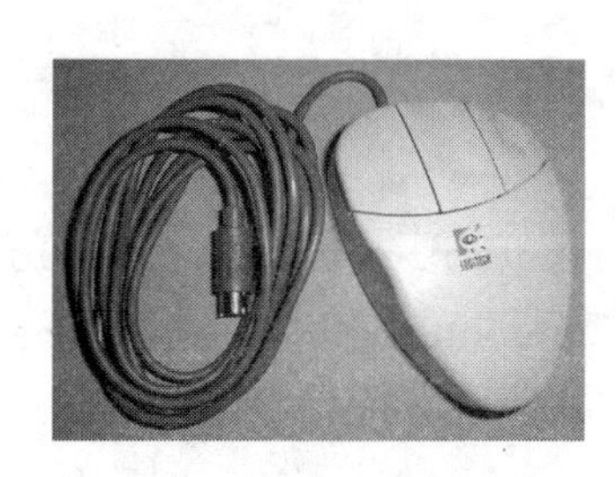

Fig. 1-1-6 Mechanical mouse

Fig. 1-1-7 Optical mouse

Fig. 1-1-8 Cordless or wireless mouse

Other pointing devices include trackball, touch screen, pointing stick, joystick, light pen, touch pad, and 3D mice.

Imaging and video input devices are used to digitize images or video from the outside world into the computer. The information can be stored in a multitude of formats depending on the user's requirement. Common types include: digital camera, video camera, webcams, image scanner, fingerprint scanner, 3D scanner, barcode reader, and laser rangefinder. Scanners allow users to input written documents, pictures, and other images into a computer system. Most scanners are of the flatbed, sheetfed, drum or handheld type (Fig. 1-1-9), among which flatbed and handheld are the most popular. Most scanners come up with optical character recognition (OCR) software that enables a computer system to edit the scanned documents. The best known uses of OCR are bar code readers used to record purchases at retail stores (Fig. 1-1-10). Famous brands of Scanners are HP, EPSON, UMAX, Canon, etc.

Audio input devices mainly refer to voice and music input devices. Voice input is usually done by a microphone and appropriate software, such as IBM ViaVoice or Dragon Naturally Speaking. Music can be input into a PC through MP3 players or MP4 players, which are also output devices.

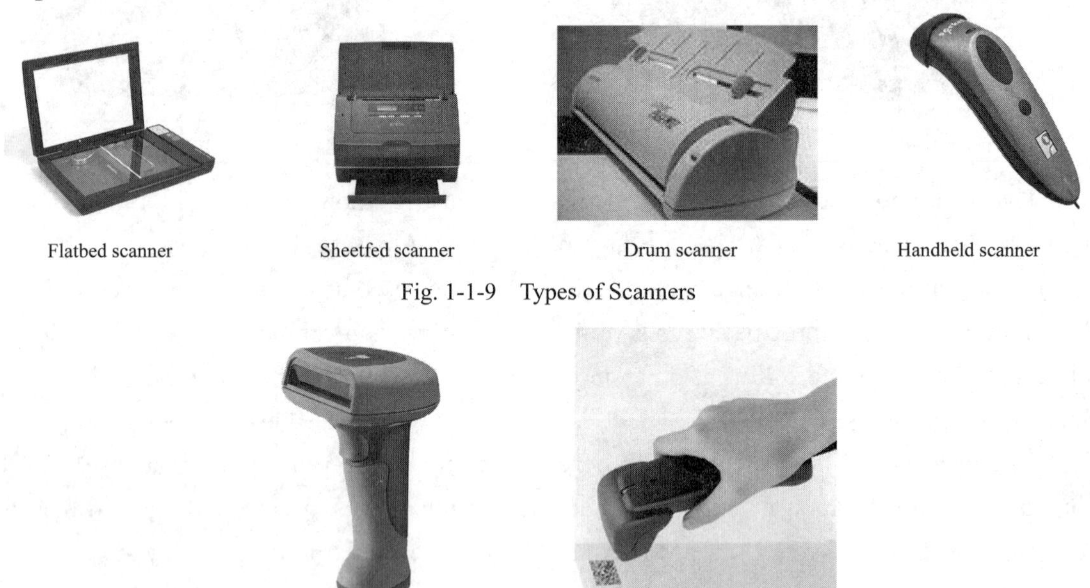

Flatbed scanner　Sheetfed scanner　Drum scanner　Handheld scanner

Fig. 1-1-9　Types of Scanners

Fig. 1-1-10　Bar code readers

Output Devices

Output devices produce results in either soft-copy or hard-copy form. Soft-copy refers to the image output on a monitor. Hard-copy refers to information output on paper — often by a printer. So the monitor and the printer are the two most common output devices.

A monitor is a hardware with a television-like viewing screen. It shows text and graphic images to the computer users, using cathode ray tube (CRT) (Fig. 1-1-11), liquid crystal display (LCD) (Fig. 1-1-12) or other image projection technology. CRTs are cheaper and display clearer images, while LCDs are thinner and occupy smaller space. Two important indexes of monitors are size and clarity. Size is indicated by the diagonal length of the monitor's viewing area. Common sizes are 23, 24, 26, and 27 inches. Clarity is indicated by its resolution, which is measured in pixels. For a fixed size monitor, the more pixels, the clearer the images. Popular brands of monitors are SAMSUNG, Philips, ViewSonic, LG , MAYA, BenQ, FOUNDER, MAG, etc.

Fig. 1-1-11 CRT

Fig. 1-1-12 LCD

A printer transfers what you see on the monitor onto paper, using impact or non-impact printing technology. The dot-matrix printer (Fig. 1-1-13) has been a popular lowcost PC printer, which uses impact printing. The best-known non-impact printers are the ink-jet printer (Fig. 1-1-14) and laser printer (Fig. 1-1-15). They are more quiet and more reliable than impact printers. Ink-jet printers can print colorful images, but most laser printers cannot. However, laser printers can print output of higher quality and at a higher speed. Famous brands of printers are Canon, EPSON, HP, SAMSUNG, etc.

Fig. 1-1-13 Dot-matrix printer

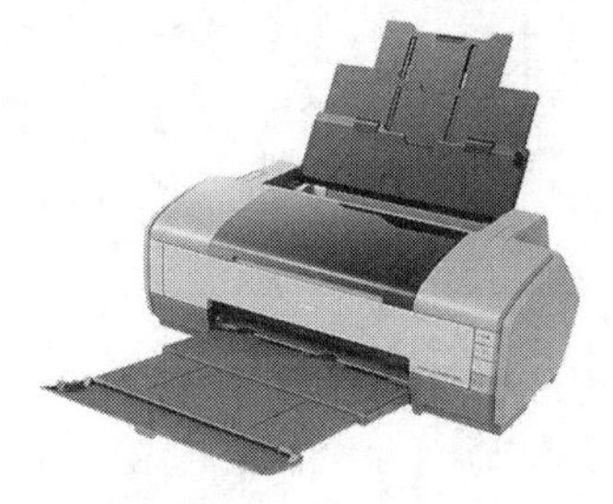

Fig. 1-1-14 Ink-jet printer

Fig. 1-1-15 Laser printer

Other output devices include: speakers and headphones, to output music or spoken voice; data and multimedia projectors, to project output for a larger audience to see.

Combination of Input and Output Devices

In order to save space or total cost, or for specialized application, many devices combine input and output devices. Common combination devices include fax machines (Fig. 1-1-16), multifunction devices (Fig. 1-1-17), Internet telephony, computer machine

(Fig. 1-1-18), etc.

Fig. 1-1-16 Fax machine

Fig. 1-1-17 Multifunction device

Fig. 1-1-18 Computer machine

Words and Expressions

instruction [ɪn'strʌkʃ(ə)n] *n.* 指令

process ['prəʊses] *v.* 处理

classify ['klæsɪfaɪ] *v.* 把……分类

function key 功能键

keypad ['kiːpæd] *n.* 键区

cursor ['kɜːsə] *n.* 光标

shortcut key 快捷键

ergonomic [ˌɜːgəʊ'nɒmɪk] *adj.* 人体工程学的

pointer ['pɒɪntə] *n.* 指针

monitor ['mɒnɪtə] *n.* 显示器

button ['bʌt(ə)n] *n.* 按钮

option ['ɒpʃ(ə)n] *n.* 选项

roll [rəʊl] *v.* 滚动

emit [ɪ'mɪt] *v.* 发射，放射

precise [prɪ'saɪs] *adj.* 精确的，准确的

infrared light [ɪnfrə'red] *n.* 红外线

trackball ['trækˌbɔl] 轨迹球

pointing ['pɔɪntɪŋ] 定位

joystick ['dʒɔɪstɪk] *n.* 操纵杆

scanner ['skænə] *n.* 扫描仪

OCR（optical character recognition） 光学字符识别

bar code reader 条形码阅读器

video camera 摄像头

webcam 网络摄像头

graphics tablet 图形输入板

microphone ['maɪkrəfəʊn] *n.* 话筒，麦克风

CRT (cathode ray tube)	阴极射线管
LCD (liquid crystal display)	液晶显示器
projection [prə'dʒekʃ(ə)n]	*n.* 投射
occupy ['ɒkjʊpaɪ]	*v.* 占据，占用
index ['ɪndeks]	*n.* 指标
clarity ['klærɪtɪ]	*n.* 清晰度
diagonal [daɪ'æg(ə)n(ə)l]	*n.* 对角线的
resolution [rezə'luːʃ(ə)n]	*n.* 分辨率
pixel ['piksəl]	*n.* 像素
transfer [træns'fɜː]	*v.* 转换，变换
impact and non-impact printing	击打式和非击打式打印
dot-matrix printer	点阵打印机
ink-jet printer	喷墨打印机
laser printer	激光打印机
speaker ['spiːkə]	*n.* 音箱
headphone ['hedfəʊn]	*n.* 耳机，头戴听筒
multimedia projector	多媒体投影仪
specialized ['spɛʃə'laɪzd]	*adj.* 专门的，专用的
combine [kəm'baɪn]	*v.* 联合，结合

Language Points

1．…the function key row **at the top of the keyboard**, the typewriter keypad, the cursor-edit keypad with arrows indicating up, down, left and right, and the numeric keypad.

主句：at the top of the keyboard 为介词短语作定语，修饰 function key row；with arrows indicating up, down, left and right 同样为介词短语作定语，修饰 cursor-edit keypad。

译文：位于键盘顶部的功能键区、打字键区、包含上下左右四个箭头移动光标的光标/编辑键区和数字键区。

2．**A mouse can have one, two or three buttons**, which are used to select command options and to control information presented on the monitor.

主句：A mouse can have one, two or three buttons. 其中，which are used to select command options and to control information presented on the monitor 为非限定性定语从句，修饰 buttons；presented on the monitor 为过去分词短语作后置定语，修饰 information.

译文：鼠标上通常有一个、两个或三个键，这些键用来选择要执行的命令和控制显示在显示器上的信息。

3．**Most scanners are of the flatbed, sheetfed, drum or handheld type**, among which flatbed and handheld are the most popular.

主句：Most scanners are of the flatbed, sheetfed, drum or handheld type. 而 among which flatbed and handheld are the most popular 为非限定性定语从句，which 指代逗号前面的 flatbed, sheetfed, drum or handheld 四种类型。

译文：扫描仪可以分为平板式、馈纸式、滚筒式和手持式，其中，以平板式和手持式两款最为流行。

4. **It shows text and graphic images to the computer users**, using cathode ray tube (CRT), liquid crystal display (LCD) or other image projection technology.

主句：It shows text and graphic images to the computer users. 代词 it 指代 monitor（显示器）；using cathode ray tube (CRT), liquid crystal display (LCD) or other image projection technology 为现在分词短语作方式状语。

译文：它使用阴极射线管（CRT）、液晶显示屏（LCD）或其他图像显像技术将文本和图形呈现给计算机用户。

5. **A printer transfers what you see on the monitor onto paper**, using impact or non-impact printing technology.

主句：A printer transfers what you see on the monitor onto paper. what you see on the monitor 为名词性从句，作 transfer 的宾语，what=the thing that；using impact or non-impact printing technology 为分词短语作状语，表示方式方法。

译文：打印机是使用击打式打印技术或非击打式打印技术将显示器上的内容转移到纸上的设备。

6. **Other output devices include: speakers and headphones**, to output music or spoken voice; **data and multimedia projectors**, to project output for a larger audience to see.

主句：Other output devices include：speakers and headphones, data and multimedia projectors, and voice-output systems。两个不定式结构 to output music or spoken voice 和 to project output for a larger audience to see 作定语，分别修饰 speakers and headphones 和 data and multimedia projectors.

译文：其他输出设备还包括用于输出音乐和声音的音箱和耳机，为更大批观众呈现清晰信息的数据和多媒体投影仪。

1.2 Motherboard and CPU

This section presents some important components of computer hardware. It is mainly about motherboard, CPU and memory.

Motherboard

The motherboard (Fig. 1-2-1) is the communications web for the entire computer system. It acts as a hub of communications, and every component of the system unit connects directly to it. Components' communications are thus allowed. The motherboard is so important that input and output devices such as monitor, keyboard, and printer can not communicate with the system unit without it.

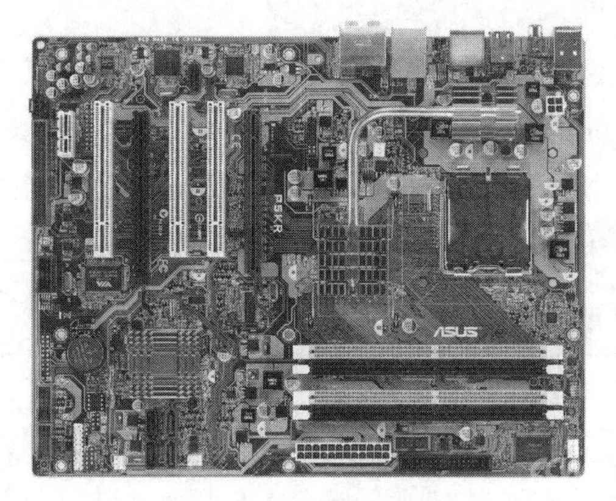

Fig. 1-2-1 Motherboard

The motherboard is located at the bottom of the desktop case, on which there are several sockets. A variety of cards, including sound card and video card, can then plug into these sockets. Cards are electronic parts made up of chips, and chips are circuit boards. The sound card is for operating the computer's sound, while the video card is for displaying graphics on the monitor.

CPU

The central processing unit (CPU) (Fig. 1-2-2) is the "brain" of the computer. It is contained on a single chip called the microprocessor and the microprocessor is often contained within a cartridge that plugs into the motherboard. CPU reads and interprets software instructions and coordinates the processing activities that must take place. Every CPU comes with a unique set of operations such as ADD, STORE, or LOAD that represent the processor's instruction set. The design of CPU affects the computer's processing power and its speed as well as the amount of main memory it can use effectively. If the CPU in your computer is designed very well, you can fulfill some tasks that are highly sophisticated in a very short time.

Fig. 1-2-2 CPU

CPU consists of two functional units — the control unit and the arithmetic-logic unit.

• Control unit

The control unit tells the rest of the computer system how to carry out a program's instructions. Under the control of this unit, programs and data are input from input devices and stored automatically and temporarily in memory, then carried out. Finally the outcome is output and printed. A program is made up of a series of instructions. The process of performing a program is the process of performing a series of instructions in a certain order. The control unit creates a series of control signals, controls the obtaining of instructions from memory, analyzes their functions, fulfills their operations, and then determines the address of the next instruction.

This step-by-step operation is repeated over and over again at an awesome speed till the program is performed.

- Arithmetic-logic unit

The arithmetic-logic unit, usually called the ALU, performs two types of operations—arithmetic and logical. Arithmetic operations perform the fundamental math operations according to arithmetic rules, such as adding, subtracting, multiplying, dividing, and evaluating absolute value and so on. Logic operations are composed of comparisons. For example, two pieces of data are compared to see whether one is equal to (=), less than (<), or greater than (>) the other. If they are equal, processing will continue; if they are not equal, processing will stop. In the computer, complex operations are often resolved into a series of arithmetic operations and logic operations. The two pieces of data in operation are called source operation data and usually stored in memory. The operation results can be stored in memory, too.

Memory

Memory is an area that holds programs processed presently and data (including the results of the operation) used by programs. When the computer runs an instruction, first, it should fetch the instruction from memory, and then execute it. If the computer needs to fetch data from memory again, it is necessary for it to visit memory once more. Therefore, the speed of memory affects directly the speed of the computer.

Like CPU, memory is contained on chips connected to the motherboard. There are three well-known types of memory chips (Fig. 1-2-3): random-access memory (RAM), read-only memory (ROM), and complementary metal-oxide semiconductor (CMOS).

Fig. 1-2-3 Memory chip

- RAM

When we speak of computer memory, we usually refer to random-access memory (RAM), which is the most widely used type. It holds the program and data that CPU is presently processing. Everything in RAM will be lost when there is a power failure or when the computer is turned off, so we say RAM is temporary or volatile storage. For this reason, we'd better store our hard work every few minutes when we are working on a document.

RAM falls into two kinds—static random-access memory(SRAM) and dynamic random-access memory (DRAM). The contents in SRAM can stay for a long time if there is no power failure, while the contents in DRAM will be lost automatically after a certain time (such as after some milliseconds) even if there is no power failure. Therefore, compared with DRAM, SRAM is more convenient and simpler to use; in addition, its speed is higher. However, SRAM has a low capacity.

- ROM

Read-only memory (ROM) holds programs. These programs are built into ROM chips when the chips are made, which means that these programs are fixed, users can use the programs and can not change them. Computers can read programs on ROM chips and can not write any information in ROM. That is why this kind of memory is called read-only memory.

Programs in ROM are some special instructions. For example, when we press the button "POWER", the computer can be started automatically; when we type in letters, they can be shown on the screen. They are all the results of instructions in ROM. It can be seen that these instructions are for computer operations.

- CMOS

A complementary metal-oxide semiconductor (CMOS) chip supplies a computer system with flexibility and expandability. Information in it concerns some essential instructions that are required whenever we turn on the computer. Differing from RAM, CMOS is powered by a battery and when there is a power failure or the computer is turned off, its contents can not be lost. Unlike fixed programs in ROM, the contents in CMOS can be changed.

Words and Expressions

plug [plʌg]	*n.* 塞子，（电）插头
socket ['sɒkɪt]	*n.* 孔，插座
graphic ['græfɪk]	*adj.* 绘画似的，图解的
cartridge ['kɑːtrɪdʒ]	*n.* 夹头，盒
sophisticated [sə'fɪstɪkeɪtɪd]	*adj.* 复杂的
volatile ['vɒlətaɪl]	*adj.* 短暂的，非永久性的
static ['stætɪk]	*adj.* 静态的，静力的
dynamic [daɪ'næmɪk]	*adj.* 动态的，动力的
awesome ['ɔːs(ə)m]	*adj.* 可怕的，令人惊叹的
resolve [rɪ'zɒlv]	*v.* 解析，使归结为(into)
execute ['eksɪkjuːt]	*vt.* 实施，执行
millisecond ['mɪlɪsek(ə)nd]	*n.* 毫秒
capacity [kə'pæsɪtɪ]	*n.* 容量，能力，性能
flexibility [ˌfleksɪ'bɪlɪtɪ]	*n.* 机动性，适应性，灵活性
expandability [eksˌpændə'bɪlətɪ]	*n.* 伸延性，扩展性
a hub of communications	交通枢纽
control unit	控制单元
arithmetic-logic unit	算术-逻辑单元
absolute value	绝对值

Language Points

1. The motherboard is so important that input and output devices such as monitor,

keyboard, and printer can not communicate with the system unit without it.

主句：此句为 so…that…句型，意思是“如此……以致于……”，that 引导结果状语从句。such as：比如。can not…without: 没有……就不能。

译文：主板很重要，没有它，输入/输出设备，如显示器、键盘、打印机等就不能与计算机系统交流。

2．Memory is an area that holds programs processed presently and data (including the results of the operation) used by programs.

主句：that 引导定语从句，修饰先行词 area。processed presently 为过去分词短语后置作定语，修饰 programs。used by programs 为过去分词短语后置作定语，修饰 data。

译文：内存是存储正在处理的程序和程序所使用的数据（包括运算结果）的地方。

3．**These programs are built into ROM chips when the chips are made**, which means that these programs are fixed, users can use the programs and can not change them.

主句：These programs are built into ROM chips when the chips are made。which means that these programs are fixed, users can use the programs and can not change them 为非限制性定语从句，指代整个主句。when the chips are made 为主句中的时间状语从句。that these programs are fixed, users can use the programs and can not change them 在定语从句中作宾语。

译文：这些程序在制作芯片时就写入 ROM 芯片里了。也就是说，这些程序是固定的，用户只能使用却不能改变它们。

1.3 Storage Device

Memory and Storage

What is memory? What is storage? Is there any difference between them? Yes, there is. It is important to know the difference between “storage” and “memory”. When people use the term “memory” in reference to computers, they are almost always referring to the computer’s main memory (or primary memory) called random-access memory or RAM, which is comprised of chips attached to the motherboard. Memory is sometimes referred to as temporary storage because it will be lost if the power to the computer is cut off. In contrast, “storage” refers to the permanent storage available to a PC, which is also called secondary storage, usually in forms of the PC’s hard drive, CDs, etc. Storage is permanent, as data and programs are retained when the power is turned off.

Storage Device and Storage Medium

All storage systems involve two parts: a storage device and a storage medium. The storage devices such as a hard disk drive, and CD or DVD drive which are installed within the computer case, write data and programs onto or read them from storage media. A storage medium must be inserted into the storage device before data and programs are read or written.

Storage Medium—Removable and Fixed Media

In many storage systems, the medium must be inserted into the device before the computer can read from it or write to it. These are called removable-media storage systems. CDs, and DVDs are examples of removable media. On the other hand, fixed-media storage systems, such as most hard disk drive systems, seal the storage medium (the hard disk) inside the storage device (hard disk drive) and users can not remove it.

- Hard disk

Nearly every PC contains one or more hard disk drive systems designed to store your programs and much of your data. Such a system is located inside the computer case and removable.

A hard disk drive system is a sealed unit. Inside the hard disk drive case there is one or more circular metal platters, on which there are many magnetic tracks. These are the storage medium—hard disk. Each platter requires two read/write heads, one for each side. These heads are mounted on a device called an access mechanism. While the shaft, along with the platters, spins at thousands of revolutions per minute, the access mechanism moves the heads in and out between the disk surfaces to access the required data.

Hard disk (Fig. 1-3-1) generally provide higher speed, larger size and better reliability at a low cost than removable-medias devices.

Fig. 1-3-1 Hard disk

However, removable-media devices have other advantages, including the following:

Unlimited storage capacity—You can insert a new medium into the storage device once one becomes full.

Transportability—You can easily share media between computers and people.

Backup—You can make a duplicate copy of valuable data on the removable medium and store the copy away from the computer.

Security—Sensitive programs of data can be stored on removable media and stored in a secured area.

- Floppy disk

Floppy disks (Fig.1-3-2), sometimes called diskettes, are removable medium and very inexpensive. They were used as one of the principle medium of storage for personal computers.

Fig. 1-3-2 Floppy disks

There are several types of floppy disks with different capacities. The traditional floppy disk is the 1.44 MB 3.5-inch disk. It is being replaced by high capacity USB flash drives.

- USB flash drive

USB flash drives (Fig. 1-3-3) are removable and rewritable. A USB flash drive consists of a memory chip encased in a small piece of plastic with a USB connector on the front. To access the data stored in a USB flash drive, you simply need to plug the device into a computer's USB port. Compared with floppy disks, they are smaller, lighter, faster, more compact, hold more data, and more reliable. The storage capacity of a USB flash drive ranges from 8 GB to 128 GB or more. USB flash drives have become increasingly common.

Fig. 1-3-3 USB flash dives

- Optical disc (disk)

Optical discs (Fig. 1-3-4) are commonly referred to as compact discs. Optical discs use laser beams to write and read data. Their storage capacity is much higher than a floppy disk. A standard CD can store up to 650 MB of data. A DVD holds even more information than a CD, because the DVD can store information on two levels. Optical discs are widely used now. They are the standard today for software delivery, as well as commonly used for storing high-capacity music and video files.

Fig. 1-3-4 Optical discs

There are several types of CDs and DVDs :

CD-ROM (compact disk read-only memory) : You can not erase and add data to the disc. You can only read data on CD-ROM discs.

CD-R (compact disk recordable memory) : You can write data on CD-R discs. But you can only write once.

CD-RW (compact disk rewritable memory) : You can erase and rewrite data on CD-RW disks.

DVD (digital versatile disc), which also includes DVD-ROM and rewritable DVDs, refers to a high-capacity optical storage format that can hold from 4.7 GB to 17 GB. Most DVD drives can play both computer and audio CDs, but you can't play DVDs in a CD drive.

Words and Expressions

memory ['mem(ə)rɪ] *n.* 内存，存储器

storage ['stɔːrɪdʒ] *n.* 存储，存储器

random ['rændəm]	*adj.* 随机的
be comprised of	由……组成
attach [ə'tætʃ]	*v.* 固定，连接
access ['ækses]	*n. & v.* 存取，访问
refer to	*v.* 谈到，指
be referred to as	被称为
temporary ['temp(ə)rərɪ]	*adj.* 暂时的
permanent ['pɜːm(ə)nənt]	*adj.* 永久的
in contrast	相反
insert [ɪn'sɜːt]	*v.* 插入
hard drive/ hard disk drive	硬盘驱动器
floppy disk	软盘
data ['deɪtə]	*n.* 数据
program ['prəʊgræm]	*n.* 程序
device [dɪ'vaɪs]	*n.* 装置，设备
medium ['miːdɪəm]	*n.* 媒介，介质
install [ɪn'stɔl]	*n. & v.* 安装
removable [rɪ'muːvəbl]	*adj.* 可移动的
circular ['sɜːkjʊlə]	*adj.*圆形的
metal ['met(ə)l]	*n.* 金属
platter ['plætə]	*n.* 大浅盘，盘
track [træk]	*n.* 轨道
mount [maʊnt]	*v.* 固定
mechanism ['mek(ə)nɪz(ə)m]	*n.* 机械装置，机构
shaft [ʃɑːft]	*n.* 轴
spin [spɪn]	*v.* 旋转
revolution [revə'luːʃ(ə)n]	*n.* 转数，圈
unlimited [ʌn'lɪmɪtɪd]	*adj.* 不受限制的
transportability [trænzˌpɔtə'bɪləti]	*n.* 可移植性
backup ['bækʌp]	*n.* 备份
duplicate ['djuːplɪkeɪt]	*n.* 复制品，副本 *v.* 复制 *adj.* 复制的
USB flash drive	U 盘
encase [ɪn'keɪs]	*v.* 装入，包住
optical disc	光盘
compact [kəm'pækt]	*adj.* 密集的，密度大的
laser ['leɪzə]	*n.* 激光
delivery [dɪ'lɪv(ə)rɪ]	*n.* 发行

digital ['dɪdʒɪt(ə)l] *adj.* 数字的
versatile ['vɜːsətaɪl] *adj.* 通用的，万能的

Language Points

When people use the term “memory” in reference to computers, **they are almost always referring to the computer’s main memory** (or primary memory) called random access memory or RAM, which is comprised of chips attached to the motherboard.

主句：they are almost always referring to the computer’s main memory。When 引导时间状语从句。过去分词短语 called random access memory or RAM 作定语修饰 computer’s main memory。Which 引导的非限定性定语从句修饰 RAM。attached to the motherboard 也是过去分词短语作定语修饰 chips。

译成：当人们谈及计算机用到“内存”这个术语时，他们通常指随机存储器 RAM，它由固定在主板上的芯片构成。

Exercises

I. Complete the following dialogue.

The problem is solved!

Peter had read the materials, and in order to buy a computer, he met Richard again. Complete their dialogues by translating the Chinese into English orally.

Dialogue

Jim: Hi. Thanks for your coming to help me.

Richard: I’ll be glad to help you. ________(1)________（现在你对计算机有所了解了吧？）

Jim: Yes. I have decided to ________(2)________（买一台组装机）. First of all, it can be very cheap. On the other hand, you know, I ________(3)________（以……为专业）business trade and know little about computer. But I think, it’s a great opportunity for me to learn about a computer when buying a DIY.

Richard: You mean you want to learn to ________(4)________（组装一台计算机）?

Jim: Yes. Could you give me some suggestions?

Richard: Sure!________(5)________（依我看）, I suggest you dispose your computer as followings: [Please give your own answer to (6)～(13)]

Motherboard: ________(6)________

CPU (central processing unit): ________(7)________

Memory: ________(8)________

Hard disk: ________(9)________

Video card: ________(10)________

Disk driver: ________(11)________

Monitor: ________________(12)________________

Printer: ________________(13)________________

As for system cabinet and speaker, it's up to you.

Jim: Great! Thank you very much! I'll write them in a list. This weekend I'll go to computer market and I can't wait to buy a DIY.

Richard: If you have any other questions, please don't hesitate to contact with me.

Jim: OK. Thanks again.

Richard: You are welcome.

II. Look at the following illustrations and label them correctly.

1.  ________________ 2.  ________________ 3. ________________

4. 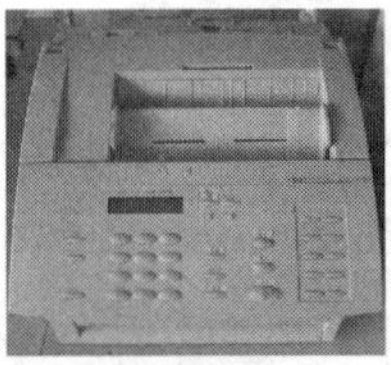 ________________ 5.  ________________ 6. ________________

7.  ________________ 8.  ________________ 9. ________________

III. Divide the following words into categories.

keyboard, monitor, CPU, scanner, ROM, RAM, hard disk, mouse, microphone, printer, sound box, CD, DVD, speaker

Input devices: ________________________________

Output devices: ________________________________

Storage units: ______________________________

Processor units: ______________________________

IV. Put the following terms into Chinese.

memory
storage
random access memory (RAM)
high resolution
input and output devices
liquid crystal display (LCD)
absolute value
ROM
CPU
arithmetic-logic unit
control unit
USB flash drive

V. Translate the following sentences into Chinese.

1. A computer is a fast and accurate system that is organized to accept, store and process data, and produce results under the direction of a stored program.

2. When people use the term “memory” in reference to computers, they are almost always referring to the computer’s main memory (or primary memory) called random access memory or RAM, which is comprised of chips attached to the motherboard.

3. Inside the hard disk drive case there is one or more circular metal platters, on which there are many tracks.

4. Clarity is indicated by its resolution, which is measured in pixels. For a fixed size monitor, the more pixels, the clearer the images.

5. A monitor is a hardware with a television-like viewing screen.

6. Input devices translate data and program instructions that people understand into a form that computers can process.

7. Most scanners are of the flatbed, sheetfed, drum or handheld type, among which flatbed and handheld are the most popular.

8. The motherboard is so important that input and output devices such as monitor, keyboard, and printer can not communicate with the system unit without it.

9. CPU reads and interprets software instructions and coordinates the processing activities that must take place.

10. RAM holds the program and data that the CPU is presently processing.

11. The process of performing a program is the process of performing a series of instructions in a certain order.

12. The sound card is for operating the computer's sound, while the video card is for displaying graphics on the monitor.

VI. Do you know the answers?

1. How many categories can computer hardware be divided into ?
2. How do people and computer communicate?
3. How many types can input devices be classified into?
4. What are the two most common output devices?
5. Could you list some combinations of input and output devices?
6. What advantages do hard disk drive system and removable-media devices each have?
7. What is the difference between control unit and arithmetic-logic unit?
8. What is the function of memory?
9. Do you know the main types of memory? What are they?
10. Is there any difference between memory and storage?

Chapter 2 Computer Software

Reading

Dialogue

What can I do?

Background: *Richard is a computer beginner. He wants to buy a personal computer. Now, he gets to a computer shop. The following is the conversation between an assistant and him.*

Assistant: Good morning. What can I do for you?

Richard: Good morning. I'd like to buy a personal computer.

Assistant: What kind of computer do you want?

Richard: I'm not sure. But my friend recommends Dell PC to me. What do you think of it?

Assistant: Dell PC is very good. Please come here and look at this new model. Looking fine, right? And it sells well recently.

Richard: OK. But what kind of operating system should I install?

Assistant: There are a variety of operating systems. What's more, PC operating system is very changeable.

Richard: Could you tell me the widely used ones?

Assistant: No problem. There are Windows, Linux, UNIX and so on. But the most widely used one is Windows and I recommend Windows line to you.

Richard: All right. I'd like to install Windows line. I remember my friend uses Windows line, too.

Assistant: OK. But there are many versions of Windows line. Now Windows 7, Windows 8, Windows 10 and Windows Vista are common ones.

Richard: Is Windows 10 the latest version?

Assistant: Yes, Windows 10 is the latest, but Windows 7 is also popular with users.

Richard: Then I'd like to take Windows 10. I think the later released, the easier to use.

Assistant: There is something in what you said.

Richard: But even if it is very easy, I don't know how to use it now.

Assistant: You can refer to a book on Windows 10. I am sure you will be familiar with it after practice.

Richard: Really? Now I can't wait to study it.

Assistant: Then this way, please.

Here is the material:

A computer is an inanimate device that has no intelligence of its own. It must be supplied with instructions so that it knows how to perform tasks. These instructions are called software. Software is another name for programs. A computer can do nothing if there is no software to feed it. Computer software can generally be divided into two basic kinds: system software and application software. We can think of system software as the kind the computer uses. Think of application software as the kind we use.

2.1 System Software

Introduction to System Software

System software refers to programs that are designed to allow the computer to manage its own internal resources. This software runs the basic operations and tells computer hardware what to do and how and when to do it. System software is not a single program. It is a collection or a system of programs that handle hundreds of technical details with little or no user intervention. System software is made up of four kinds of programs: operating system, utilities, device drivers and language translators.

- Operating system

People always hope to make full use of all the resources of computer system, and enhance the functions of computer system and provide a convenient working environment for users. Operating systems have developed just for these purposes. The operating system is the most important system software and the essential part in computer system. It interacts with the application software and the computer. The main functions of it are job management, memory management and device management. There are many operating systems, among which the most widely used are Windows, UNIX, and Linux. Some of them will be discussed later.

- Utilities

Utilities perform specific tasks (Fig. 2-1-1) connected with managing computer resources. They are specialized programs designed to make computing easier. When a computer is working, any problem may happen. For example, viruses may destroy a system so that computers can not run; hard disks may crash so that information stored in them can be lost, and so on. These kinds of problems can be solved with utilities.

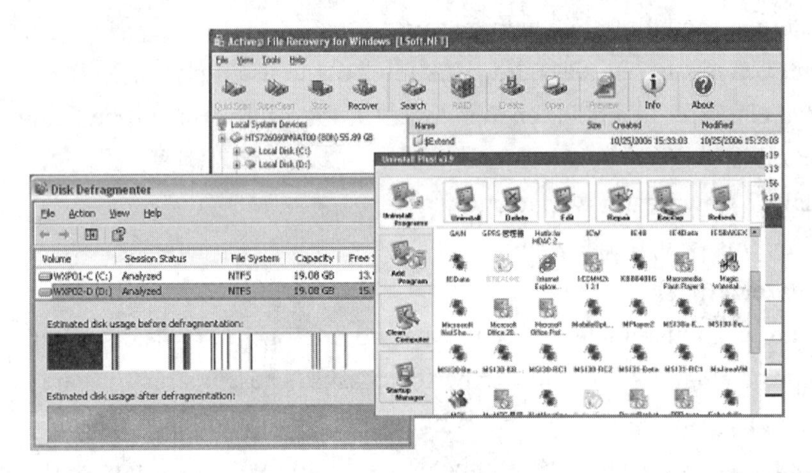

Fig. 2-1-1 Utilities to system software

There are different utility programs to execute specific tasks. The following are commonly used ones.

Backup programs are designed to copy files selected in case the original ones are lost or damaged when there is a disk failure. The software will compress the data to take up the least amount of space.

File compression programs squeeze out the slack space generated by the formatting schemes.

Antivirus programs protect computer system from being damaged by viruses invading in computers. These programs are must-have programs because viruses are spreading now and then.

Disk defragmenter is to locate and eliminate unnecessary fragments and rearrange files and unused disk space to optimize operations.

Uninstall programs are used to remove unneeded programs and related files from hard disks completely and safely.

File recovery programs are used to recover deleted or damaged files.

- Device drivers

A device driver is a special program that works with the operating system to allow communication between the device and the rest of the computer system. Every device, such as a printer, sound card, video card, mouse, must have its own driver, so the operating system can load them into memory when the computer system is started. A device driver translates commands from the operating system or user into commands understood by the device it interfaces with.

Whenever a new device is added to a computer system, a new device driver must be installed before the device can be used. If we have not a particular driver in hand, we can unload it from the manufacturer's website.

- Language translators

Both high-level languages and low-level languages are called programming languages. Low-level languages are also called machine languages consisting of zeros and ones. They are the most basic type of programming languages and can be understood directly by computers. High-level languages, such as C, C++, Pascal, are written by programmers. High-level languages must be first translated into a machine language before they can be understood and processed by computers. Language translators are software that translates high-level languages into machine languages. There are three types of language translators: assembler, compiler and interpreter.

Words and Expressions

inanimate [ɪn'ænimət] *adj.* 无生命的，死的
instruction [ɪn'stʌkʃ(ə)n] *n.* 命令
intervention [ˌɪntə'venʃ(ə)n] *n.* 介入，干涉
utility [juː'tɪlətɪ] *n.* 应用程序
interact [ˌɪntər'ækt] *vi.* 相互作用，相互影响
enhance [ˌɪn'hɑːns] *vt.* 增加，提高，增进
essential [ɪ'senʃ(ə)l] *adj.* 不可少的，必要的
execute ['eksɪkjuːt] *vt.* 执行
unload [ʌn'ləʊd] *vt.* 卸载
assembler [ə'semblə] *n.* 汇编程序
compiler [kəm'paɪlə] *n.* 编译程序
interpreter [ɪn'tɜːprɪtə] *n.* 解释程序
system software 系统软件
application software 应用软件
operating system 操作系统
device driver 设备驱动程序
language translator 语言翻译器
backup program 备份软件
file compression program 文件压缩程序
antivirus ['æntɪvaɪrəs] *n.* 杀毒软件
disk defragmenter 磁盘碎片整理
uninstall program 卸载软件
file recovery program 文件恢复软件

Language Points

1. **<u>There are many operating systems</u>**, among which the most widely used are Windows, UNIX, and Linux.

主句：which 引导非限制性定语从句，修饰先行词 operating systems。among 是介词，which 在定语从句中作 among 的宾语。

译文：现在有很多种操作系统，应用最广泛的有 Windows, UNIX 和 Linux。

2. **<u>Backup programs are designed to copy files selected</u>** in case the original ones are lost or damaged when there is a disk failure.

主句：Backup programs are designed to copy files selected。selected 是过去分词后置作定语，修饰 files。in case 引导条件状语从句。when there is a disk failure 是条件状语从句中的时间状语从句。

译文：备份软件是为了防止原始文件在磁盘损坏时丢失或受损而对文件进行复制的软件。

3. A device driver translates commands from the operating system or user into commands understood by the device it interfaces with.

主句：translate sth. into sth. from the operating system or user 修饰 commands。understood by the device it interfaces with 是过去分词短语后置作定语，修饰它前面的 commands。it interfaces with 是定语从句，修饰先行词 the device。

译文：设备驱动程序将操作系统或用户的命令翻译成它所连接的设备能够理解的命令。

Windows Operating System

There have been many different versions of Microsoft's Windows operating systems in the last several years. The earliest version of Windows is Windows Version 1.01. It is very simple. Then there are Windows 3.x, Windows 9.x, Windows NT, Windows 2000, Windows XP, Windows Vista, Windows 7, Windows 8, Windows 10 and so on. Several versions will be presented here.

Windows version 3.x was released in the 1980s and early 1990s. Windows 3.x stands for the version number of the software, such as Windows 3.0, 3.1 and so on. It provided a graphical user interface for DOS computers. Instead of the DOS command line, Windows 3.x used a system of menus, windows, and icons.

In 1995, Microsoft released a new version of Windows for personal computers called Windows 95. It was designed to replace DOS and Windows 3.x. Windows 95 has a simpler graphical user interface than previous versions. The graphical user interface is not the shell, but is integrated into the operating system. Like Windows 3.x, Windows 95 uses windows and a desktop. It supports e-mail, fax transmission, multimedia, long filenames, and Plug & Play, which make the process of installing new hardware easier. Windows 98 is similar to Windows 95, but enhances several capabilities. Windows 98 adds a higher degree of Internet integration, provides additional commands for customizing the desktop user interface, support for large hard drives, and support for both DVD disks and USB (universal serial bus).

Windows 2000 is the upgrade to Windows NT. There exist several versions of Windows 2000. In fact, Windows 2000 line operating systems consist of four separate products. They are Windows 2000 Professional, Windows 2000 Server, Windows 2000 Advanced Server and Windows 2000 Datacenter Server.

On October 22, 2009, Microsoft released Windows 7. Unlike its predecessor, Windows Vista, which introduced a large number of new features, Windows 7 was intended to be a more focused, incremental upgrade to the Windows line, with the goal of being compatible with applications and hardware which Windows Vista was not at the time. Windows 7 has multi-touch support, a redesigned Windows shell with a new taskbar, referred to as the Superbar, a home networking system called HomeGroup, and performance improvements.

On 29 February 2012, Microsoft released Windows 8 Consumer Preview, the beta version of Windows 8. For the first time since Windows 95, the Start button is no longer available on the taskbar, though the Start screen is still triggered by clicking the bottom-left corner of the screen and by clicking Start in the Charm. Windows president said more than 100,000 changes had been made since the developer version went public. In the first day of its release, Windows 8 Consumer Preview was downloaded over one million times.

Windows 10 was released on July 29, 2015. The new product not only can be installed in personal computers but also in smartphones and tablet. This version restores the Start menu that users know and love, which also allows them to easily find their files, applications. Windows 10 start and reset quickly and help the batteries to last longer. It is designed to be compatible with all devices and Windows applications and therefore it can be used both with a keyboard or/and a touch screen.

Words and Expressions

icon ['aɪkɒn]	*n.* 图标
megabyte ['megəbaɪt]	*n.* 兆字节
integrate ['ɪntɪgreɪt]	*vt.* 使结合(with)，使并入(into)
transmission [trænz'miʃ(ə)n]	*n.* 传递，传输
upgrade ['ʌpgreɪd]	*vi.* 升级
flagship ['flægʃɪp]	*n.* 旗舰
laptop ['læptɒp]	*n.* 便携式计算机
stability [stə'bɪlɪti]	*n.* 稳定性
adaptability [əˌdæptə'bɪlɪti]	*n.* 适应性
customize [kʌst əmaɪz]	*vt.* 定制，按规格改制
professional [prə'feʃ(ə)n(ə)l]	*adj.* 专业的，职业的
administrator [əd'mɪnɪstreɪtə]	*n.* 管理人
clustering ['klʌstərɪŋ]	*n.* 聚集，群集

predecessor ['pri:disesə] n. 前任，前辈
incremental [ˌɪnkrɪ'mentəl] adj. 增加的，增值的
compatible [kəm'pætɪb(ə)l] adj. 兼容的，能共处的
graphical user interface 图形用户界面
Plug & Play 即插即用
DVD (digital video disk) 数字化视频光盘
USB (universal serial bus) 通用串行总线
load balancing 负载均衡
datacenter server 数据中心服务器
command line 命令行
Consumer Preview 用户体验版

UNIX and Linux Operating Systems

- UNIX operating system

UNIX operating system was originally developed by Dennis Ritchie and Ken Thompson at AT&T Bell Laboratories. Since its development in the early 1970s, UNIX has been enhanced by many individuals, especially by computer scientists at the University of California, Berkeley. There are a large number of different versions of UNIX.

It is well known that Windows operating system is designed for Intel-type chips and Mac operating system is designed for Power PC chips. However, UNIX operating system is very flexible. It supports different processors. It can be used on a wide variety of computer systems, ranging from personal computers to mainframes. UNIX can also easily integrate many devices from different producers through network connection.

UNIX is a multi-user, multitasking operating system. It allows from one to hundreds of users to run different programs at the same time. It also allows each user to run more than one program simultaneously.

- Linux operating system

As a young student at the University of Helsinki in Finland, Linus Torvalds developed Linux in 1991. It is a free UNIX clone that supports a wide range of software. Linus has provided the operating system free to others and has encouraged further development. Because the source code for the entire Linux operating system is freely available, anyone can modify the program to improve it or customize it to a particular application.

Linux operating system contains all of the features that people would expect in UNIX, or any other operating system, such as multi-user, multitasking, the world's fastest TCP/IP drivers and so on. It supports 32-bit and 64-bit multitasking. This operating system also includes advanced networking capabilities. Networking support in Linux is superior to most other operating systems.

Linux operating system has a bright future. Apart from the fact that it is freely distributed, Linux's functionality, adaptability and robustness have made it the main alternative for UNIX and Microsoft operating systems.

Words and Expressions

individual [ɪndɪvɪdjʊ(ə)l]	*adj.* 个人的，个体的，个别的
flexible ['fleksəbl]	*adj.* 灵活的，可塑造的
mainframe ['meɪnfreɪm]	*n.* 大型机
simultaneously [sɪml'teɪnɪəsli]	*adv.* 同时地
clone [kləʊn]	*n.* 克隆
modify ['mɒdɪfaɪ]	*vt.* 变更，修改
capability [keɪpə'bɪlɪti]	*n.* 能力，性能
distribute [dɪs'trɪbjuːt]	*vt.* 分布，散布
functionality [fʌŋkəʃə'næləti]	*n.* 功能性
robustness [rəʊ'bʌstnɪs]	*n.* 健壮性，稳健性
alternative [ɔːl'tɜːnətɪv]	*adj.* 两者择一的，选择的
range from	从……延伸到
source code	源代码

Language Points

Apart from the fact that it is freely distributed, Linux's functionality, adaptability and robustness have made it the main alternative for UNIX and Microsoft operating systems.

主句：that it is freely distributed 作 the fact 的同位语从句。it 为代词，作宾语，代 Linux。the main alternative 作 it 这个宾语的补足语。

译文：它不仅免费发布，而且它的功能性、适应性和稳健性都使其成为 UNIX 和微软操作系统的主要替代品。

2.2 Application Software

Introduction to Application Software

Nowadays, there are thousands of software products for computer users to choose. These software products conduct a wide range of different tasks. Computer users can use them to write letters, send E-mails, organize finances, create graphs, learn foreign languages, make presentations, and do many other applications. These software products perform specific tasks. We call this type of software application software.

Application software can be divided into two categories. One is general-purpose applications, and the other is special-purpose applications. General-purpose applications include word processors, spreadsheets, presentation graphics and so on. Special-purpose applications include many other programs. These programs focus on specific occupations. Multimedia and Web authoring are well-known special-purpose applications.

When we use application software, we must know the following features.

A window (Fig. 2-2-1) is a rectangular area that contains a document. One or more documents can be edited at the same time. Each document appears in a separate window. This is particularly valuable when working on a large project that consists of several different files. Windows can be resized, moved and closed.

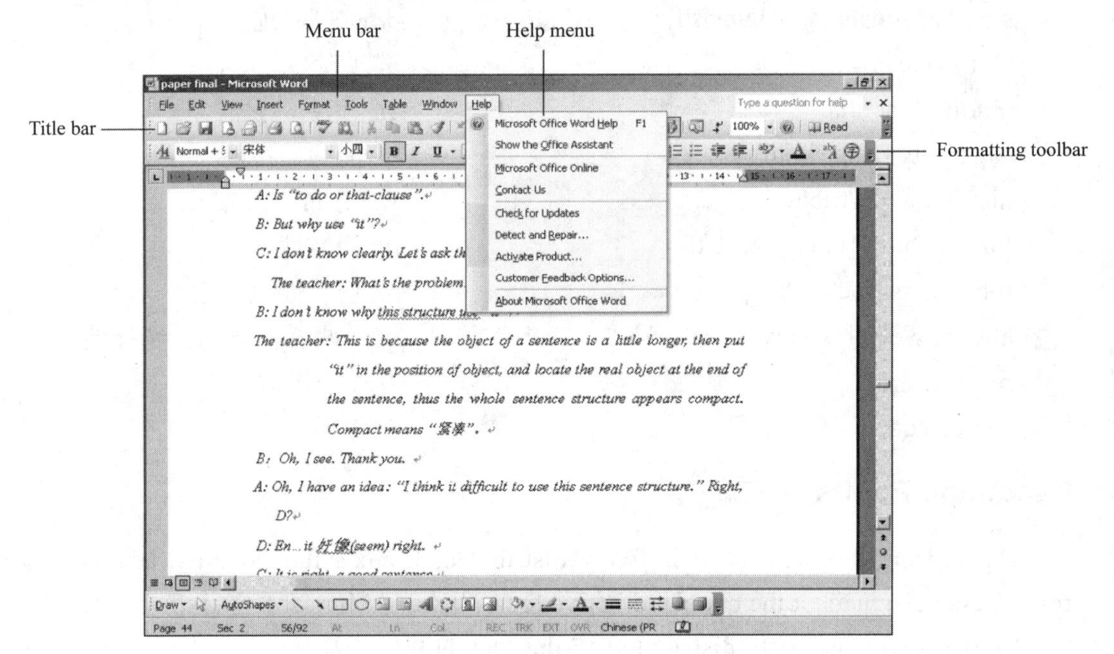

Fig. 2-2-1 A window

At the top of the computer screen, there is a menu bar. It displays menus, which consist of commands. When we select a menu and click it, we will see a pull-down menu. The options in a pull-down menu are the commands related to the menu.

On the menu bar, there is a command called help. The help menu provides us with a variety of help features. It offers explanations of how to fulfill various tasks.

Toolbars are located below the menu bar. They are buttons that are used frequently. Toolbars include the standard toolbar and the formatting toolbar. They all help us use some commands quickly.

Sometimes there appears a dialog box on the screen. It is used to collect information from the user and to present helpful messages.

When we type information, we need spaces. Text boxes provide spaces for us to enter a number, the name of a particular document or folder. After entering the information into a text box, we need to press the key Tab. Then we will move to the next dialog box.

Words and Expressions

general-purpose	通用的
special-purpose	专用的
rectangular [rek'tæŋgjʊlə]	*adj.* 矩形的，长方形的
menu bar	菜单栏
pull-down menu	下拉菜单
option ['ɒpʃən]	*n.* 选择，选择权
fulfill [fʊl'fɪl]	*vt.* 执行，完成
toolbar [tu:l bɑ:(r)]	工具栏
standard toolbar	标准工具栏
formatting toolbar	格式工具栏
dialog box	对话框
text box	文本框
folder ['fəʊldə]	*n.* 文件夹

Word Processors

Word processing software allows people to create many types of personal and business communications, including reports, announcements, letters, memos, manuscripts, as well as other forms of written documents. People can create, edit, save, and print them. Word processors are one of the most flexible and widely used application software tools.

Today, the most widely used word processing programs are Microsoft Word for the PC, Corel WordPerfect for the PC, and WordPerfect for the Mac.

The following paragraphs deal with the features of Word processing software.

- Creating documents

Using Word processing software to create a document is very easy. We just use keyboard to enter the contents of text. All Word processors have a feature called word wrap. When we type, word wrap automatically begins a new line of text once the current line is full. However, if we want to begin a new paragraph or leave a blank line, it is necessary for us to press the key Enter.

- Editing documents

When we edit documents, the following features of word processors will be employed when needed.

Insert and delete: inserting means adding new text into the document. When we do the act of inserting, first, we should move the curser to the position where we need to insert text and then begin typing. Deleting text has two ways. One is to use the key Delete. When we use this key, we should place the curser before text we need to delete and then press the key. The other way is to use the key Backspace. We need to move the curser after text we need to delete and then press Backspace.

Undelete: the Undelete command means that after we delete a block of text, we change our mind and restore what we delete. If we do several Delete command continuously, we can click the Undelete command continuously and restore all the deleted text.

Spelling and grammar checkers: these features help us write better. When we enter text, spelling errors are automatically identified. A list of similar words is advised for us to choose from. Some programs have Auto Correct function, which can correct some common mistakes such as using "and" instead of transposed letters "adn". However, if the words we enter are not in the program's dictionary, these words are also marked spelling errors, such as some technical terms or proper names. Grammar checker helps us identify grammar errors, such as subject-verb agreement, incomplete sentences, punctuation, and capitalization problems. These grammar errors can be corrected by selecting from a list of proposed corrections. The thesaurus is word processors' another feature. When we look up a word, it provides us with related words.

Cut, copy and paste: if we want to move a paragraph or block of text from one place to another location, we simply use the Cut or Copy command. First, select the part of text we want to move and select the Cut or Copy command. Then place the curser to the position where we want to move, and use the Paste command to insert it.

Search and replace: the Search command, also called the Find command, helps us locate any character, word or phrase in documents, while the Replace command helps us replace the located text with other text we specify. For example, we can quickly find each occurrence of the word "White" in a document using the Search command, and click the Replace command to replace it with another word "Black".

- Formatting documents

Formatting documents means changing the appearance of a document. It includes many choices.

First, we can format characters. Select the part we need to format with the mouse, and use the appropriate format. We can decide what typeface, font size and font style we wish to use. A typeface is a set of characters with a specific design, such as Times New Roman, Arial Black. It is commonly called a font. Font size refers to a character's height and width. It is commonly measured in points. For example, we can choose 14-point Arial Black as a title's typeface and font size. We can also decide whether the font should be underlined, boldface or italic.

Then we can format paragraphs (Fig. 2-2-2). Common paragraph formats include line spacing, margins, alignment, and indentation. Line spacing means whether we want the lines to be single-spaced, double-spaced or something else. Margins are the dimensions of before and after paragraphs. There are five types of paragraph alignment: justified, centered, left, right and distributed. Justified means aligning text evenly between left and right margins. Centered means that every line of paragraphs is centered on the page. Left means the left side of the text is evenly aligned, and right means the right side of the text is evenly aligned. Distributed means that characters on each line of a paragraph are arranged evenly between left and right margins. Indentation includes first line and hanging.

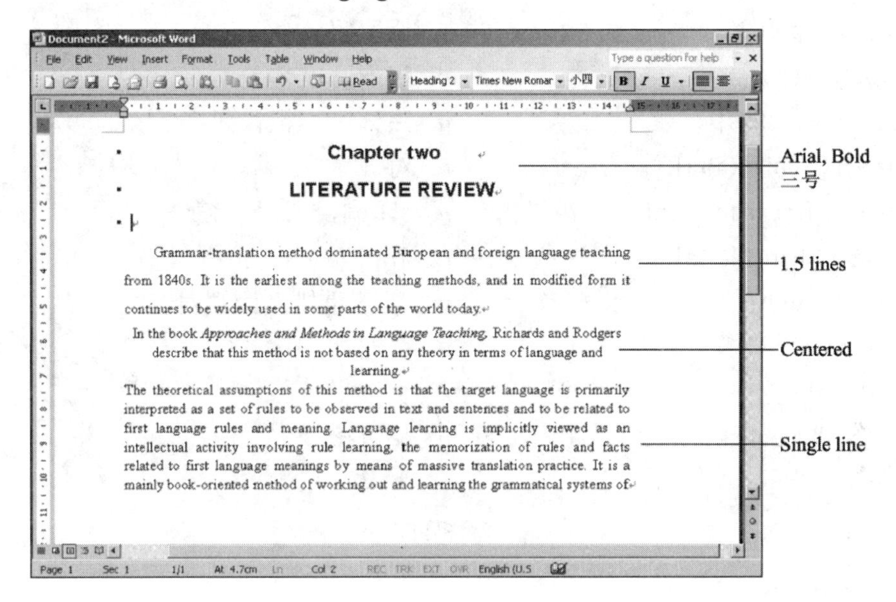

Fig. 2-2-2 Format paragraphs

Most word processing programs also have page formatting options. We can indicate page numbers at the top or bottom of the page. We can add headers and footers. A header is printed automatically at the top of every page. It is usually a date, document name or something like this. A footer is printed at the bottom of every page. We can also format documents. In fact, when we format all the pages of a document, we call it document formatting. We can apply a background to a document, specify footnotes, generate tables, and add graphics from other files.

Some word processors include these features: access Web pages and search for information, design and publish Web pages on the Internet, insert a hyperlink to a Web page in a word processing document, and retrieve pictures from other Web pages.

Words and Expressions

processor ['prəusesə]	n. 处理器
memo ['meməu]	n. [口]=memorandum 备忘录
manuscript ['mænjuskrɪpt]	n. 手稿
cursor ['kɜːsə]	n. 光标，指针
transpose [træns'pəuz]	vt. 换位，调换，颠倒
punctuation [pʌŋ(k)tʃu'eɪʃ(ə)n]	n. 标点符号
capitalization [ˌkæpɪt(ə)laɪ'zeɪʃ(ə)n]	n. 大写
thesaurus [θɪ'sɔːrəs]	n.（知识的）宝库，词典
locate [lə(u)'keɪt]	vt. 确定……的地点

occurrence [ə'kʌr(ə)ns]	*n.* 发生，出现，存在
format ['fɔːmæt]	*n.* 版式，格式 *vt.* 格式化
typeface ['taɪpfeɪs]	*n.* 铅字字体
font [fɒnt]	*n.* 字形，字体
boldface ['bəʊldfeɪs]	*n.* 黑体字，粗体黑字
alignment [ə'laɪnm(ə)nt]	*n.* 对齐
indentation [ɪnden'teɪʃ(ə)n]	*n.*（印刷，书写）缩进
dimension [dɪ'menʃ(ə)n]	*n.* [pl.] 大小，范围
justify ['dʒʌstɪfaɪ]	*vt.* 段落重排，两端对齐
header ['hedə]	*n.* 页眉
footer ['fʊtə]	*n.* 页脚
hyperlink ['haɪpəlɪŋk]	*n.* 超链接
retrieve [rɪ'triːv]	*vt.* 检索
word wrap	换行

Language Points

The Undelete command means that after we delete a block of text, we change our mind and restore what we delete.

主句：从句 that after we delete a block of text, we change our mind and restore what we delete 作 means 的宾语。after we delete a block of text 是宾语从句中的时间状语从句。what we delete 是 what 引导的名词性从句，作 restore 的宾语。

译文：撤销删除命令指的是在删除了文本的一部分内容之后，又改变主意，可以恢复删除。

Spreadsheets

Spreadsheet software is widely used by people in nearly every profession. People use it to organize, analyze numeric data and create graphs. Today, the most widely used spreadsheet programs are Microsoft Excel, Lotus 1-2-3 and Corel Quattro Pro.

The following will discuss the features of spreadsheets.

A spreadsheet is a rectangular grid formed by rows and columns. Rows are marked by numbers and columns are marked by letters. A cell is created by the intersection of a row and a column. A number, word or phrase, or formula is stored in each cell. Numbers and formulas are called numeric entries, while words and phrases are called text entries.

Formulas are instructions for calculations. Spreadsheet programs offer a rich environment to build complex formulas. Using a few mathematical operators and rules for cell entry, we can do a lot of calculations. For example, the formula B4+B5+B6 means to add the values in B4, B5 and B6. If we enter 1, 2 and 3 into B4, B5 and B6 respectively, select the blank cell B7 and type the formula "= B4+B5+B6", press the key Enter, 6 appears in cell B7 (Fig. 2-2-3).

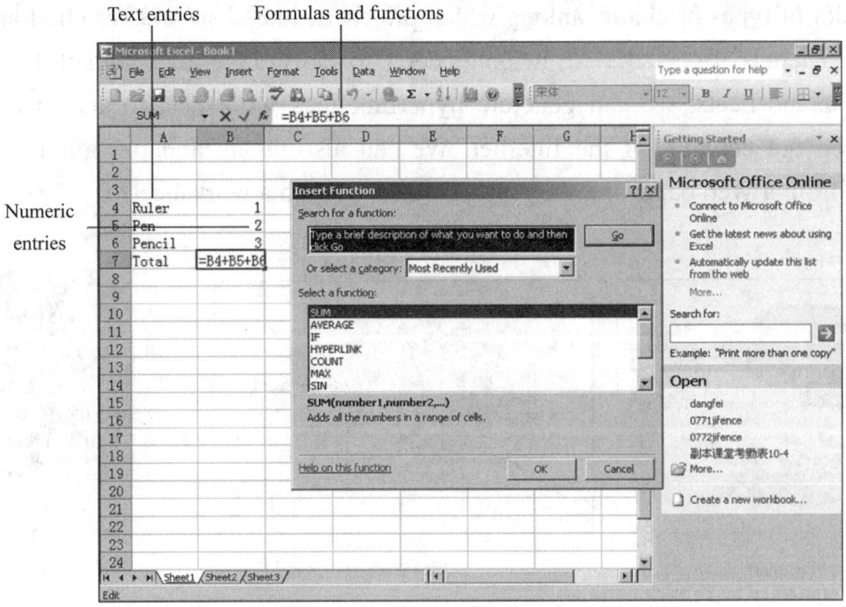

Fig. 2-2-3 The operation of sum

Functions are spreadsheet software's special tools. They can do complex calculations automatically, easily and quickly. In fact, functions are prewritten formulas that perform certain types of calculations automatically. They are shorthand versions of frequently used formulas. For example, suppose we want to add the values in the cells from A1 to A5, if we use the formula, we should enter "=A1+A2+A3+A4+A5"; if we use the function, we enter "= SUM (A1:A5)". It is clear that the SUM function is a lot shorter than the formula and makes our work much easier.

Spreadsheet software can do recalculation. This means if we change one or more numbers in a spreadsheet, all related formulas will recalculate automatically and produce a new result. For example, when we enter 10 into the cell A1 and "=A1+5" into the cell A2, press the key Enter, 15 appears in A2. If we substitute 5 for the value in A1, and then select the cell A2, the value in it becomes 10 automatically. This simple function can greatly reduce our hours for hard work.

Spreadsheet software has another feature, that is, there is an AutoSum button on the Standard toolbar. The AutoSum button means inserting the SUM function into a cell and suggests a range to sum. If the suggested range is not correct, we simply drag through the correct range, and press the key Enter. The AutoSum button also has a menu. We can find the Average, Count, Max, or Min function in it. In addition, there is a More Functions command. If we select the More Function command, the Insert Function dialog box appears, and we can access any other functions.

Another useful feature of spreadsheet software is Chart Wizard. Chart Wizard helps us to generate charts. Charts display the data in worksheets directly. They are easy to understand. There are a lot of types of charts, among which pie chart, line chart and bar chart are the most widely used. There is also stock chart helping people to analyze the stock market (Fig. 2-2-4).

Using spreadsheets, we can generate hyperlink in a worksheet to access other Office documents on the network or the Internet. We can also create and run queries to retrieve information from a Web page, and then insert it directly into a worksheet.

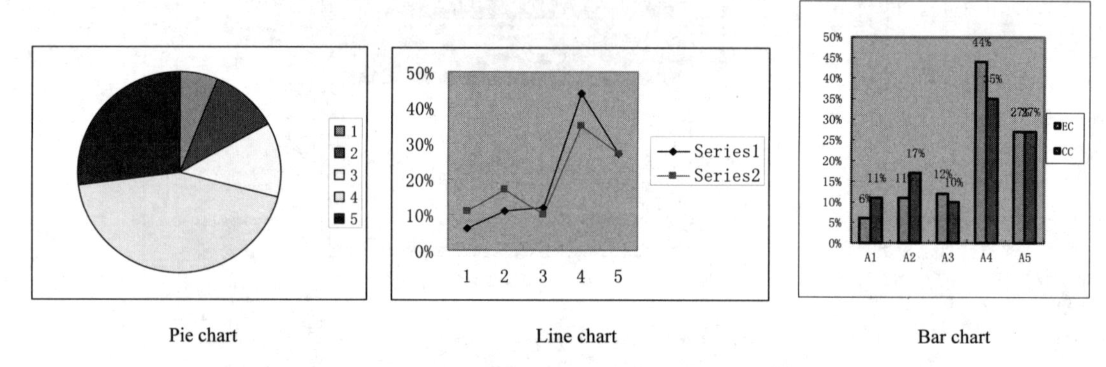

Pie chart Line chart Bar chart

Fig. 2-2-4 Types of charts

Words and Expressions

spreadsheet ['spredʃiːt]	*n.* 空白表格程序，电子表格处理软件
profession [prə'feʃ(ə)n]	*n.* 职业
numeric [njuː'merɪk]	*adj.* (=numerical) 数字的，数值的
grid [ɡrɪd]	*n.* 格栅，格子
intersection [ˌɪntə'sekʃ(ə)n]	*n.* 横断，交叉
formula ['fɔːmjʊlə]	*n.* 公式，方程式
function ['fʌŋ(k)ʃ(ə)n]	*n.* 函数
substitute ['sʌbstɪtjuːt]	*n.* 代替者
query ['kwɪəri]	*n.* 查询
entry ['entri]	*n.* 输入项，条目
respectively [rɪ'spektɪvli]	*adv.* 各自地，分别地
AutoSum button	自动求和按钮
More Functions command	其他函数命令
Chart Wizard	图表向导

Presentation Graphics

Presentation graphics mean using visual objects—graphics and information, to create presentations. Presentations are visually interesting and attractive. They are widely used in business and classrooms. Many people employ presentation graphic software to make their report, give their lectures, present their information, and so on. Today, the popular presentation graphic programs are Microsoft PowerPoint, Lotus Freelance Graphics and Aldus Persuasion.

Here we will mainly discuss the features of Microsoft PowerPoint.

When the PowerPoint window is opened, we can see a slide. This is the first slide. We call it the "Title Slide". The Title Slide contains placeholders where we can place a title and a subtitle. If we do not want to have a subtitle, we simply ignore the subtitle placeholder and it will not appear on the presentation. When we click the box "Click to add title", we can enter text (Fig. 2-2-5). To add a new slide to the presentation, we should click the "New Slide" button. Every time we click the "New Slide", we will get a default slide.

When we want to display slides, click "View" on the menu bar. We can choose "Slide Show" or "Slide Sorter". Another way is the normal view. When we open PowerPoint, we are faced with this default view. The last way is outline view. This way allows us to work with the text of our whole presentation in an outline format.

Microsoft PowerPoint has a feature, that is, it can create dynamic presentations easily. We also call it animations. They include special visual and sound effects, such as pictures, movies, and sounds. There are templates which can help us design a presentation. A template is a file that includes predefined settings. These predefined settings can be used as a pattern to create many common types of presentations.

Fig. 2-2-5 Title slide

Words and Expressions

presentation [ˌprezən'teɪʃən]	*n.* 呈现，展示，演示
visual ['vɪzʊəl]	*adj.* 视觉的，视力的，看得见的
slide [slaɪd]	*n.* 幻灯片
placeholder ['pleɪsˌhəʊldə]	*n.* 占位符

title ['taɪt(ə)l]	*n.* 标题
subtitle ['sʌbtaɪt(ə)l]	*n.* 副题，小标题
ignore [ɪg'nɔː]	*vt.* 忽视，不理睬
default [dɪ'fɔːlt]	*n.* 默认，默认值
outline ['əʊtlaɪn]	*n.* 大纲，提纲
dynamic [daɪ'næmɪk]	*adj.* 动力的，动态的
animation [ˌænɪ'meɪʃ(ə)n]	*n.* 动画
template ['templeɪt]	*n.* 模板
setting ['setɪŋ]	*n.* 设置，背景

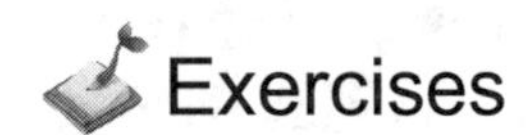

Exercises

I. Complete the following dialogue.

The problem is solved!

Two days later, Richard met his friend John. He was very excited and told John what he learned about some application software.

Dialogue

John: Hi, Richard. How is everything going with you?

Richard: Everything is terrific. John, you know I bought a new computer and I really know something about how to use it.

John: Then tell me what you know now.

Richard: I know how to ____(1)____ (创建和编辑文档) using Microsoft Word. This software is miraculous. It has ____(2)____ (拼写和语法检查) and can correct my mistakes automatically. What's more, I can ____(3)____ (删除) what I don't want, but if I change my mind, it can ____(4)____ (撤销删除) and what I want appears again. That's really convenient. In addition, it has the features of ____(5)____ (粘贴和复制).

John: Great! John. You know so much now. In fact, Microsoft Word has many other features. For example, we can ____(6)____(查找和替换) any word in a document; we can change ____(7)____(字体、字号和字形); we can even ____(8)____(格式化) paragraphs.

Richard: It seems there is still a lot for me to learn.

John: I'm not familiar with some features, either. You know I have been using computers for a long time. By the way, have you used Microsoft Excel?

Richard: I have just browsed it. I know people use it to organize, analyze numeric data and____(9)____ (创建图表).

John: Yeah. It's also easy to learn. You will find it's very useful, too.

Richard: I will use it soon. If I have any problem, can you help me?
John: Of course.

II. Match the English in column A with the Chinese in column B.

A	B
language translator	数字化视频光盘
AutoSum button	数据中心服务器
More Functions command	负载均衡
Chart Wizard	语言翻译器
DVD	图表向导
USB	自动求和按钮
load balancing	其他函数命令
datacenter server	通用串行总线

III. Put the following terms into Chinese.

system software
application software
device driver
backup program
file compression program
antivirus
disk defragmenter
uninstall program
file recovery program
graphical user interface
Plug & Play
command line
source code
word wrap

IV. Translate the following sentences into Chinese.

1. Computer software can generally be divided into two basic kinds: system software and application software.

2. System software is made up of four kinds of programs: operating system, utilities, device drivers and language translators.

3. The main functions of operating systems: job management, memory management and device management.

4. Whenever a new device is added to a computer system, a new device driver must be

installed before the device can be used.

5. High-level languages must be first translated into a machine language before they can be understood and processed by computers.

6. The graphical user interface is not the shell, but is integrated into the operating system.

7. It can be used on a wide variety of computer systems, ranging from personal computers to mainframes.

8. Word processing software allows people to create many types of personal and business communications, including reports, announcements, letters, memos, manuscripts, as well as other forms of written documents.

9. When we type, word wrap automatically begins a new line of text once the current line is full.

10. Common paragraph formats include line spacing, margins, alignment, and indentation.

11. We can apply a background to a document, specify footnotes, generate tables, and add graphics from other files.

12. The AutoSum button means inserting the SUM function into a cell and suggests a range to sum.

13. Using a few mathematical operators and rules for cell entry, we can do a lot of calculations.

14. The Title Slide contains placeholders where we can place a title and a subtitle.

15. For the first time since Windows 95, the Start button is no longer available on the taskbar, though the Start screen is still triggered by clicking the bottom-left corner of the screen and by clicking Start in the Charm.

V. Do you know the answers?

1. What is software? How many kinds is it divided into?
2. Do you know the main kinds of system software? What are they?
3. Could you list some commonly used utility programs?
4. What are the three types of language translators?
5. What are the features of Windows 10?
6. What are main functions of Word processors?
7. How many types of paragraph alignment? Please list them.
8. In a worksheet, if we want to change data into charts, what function should we use?
9. What do people employ presentation graphic software to do?
10. Do you know how to show slides?

Chapter 3 Programming Language

Reading

Dialogue

What can I do?

Background: *Richard got very good grades for all the courses after the mid-term exam. However, he heard that there would be a course about programming language next term, and it was not easy to learn. Richard wanted to do some preparations but had no idea where to start. He came to his teacher Mr. Liu for help.*

Richard: Good morning, Mr. Liu.

Mr. Liu: Morning, Richard. Come in, please.

Richard: Mr. Liu, I have some trouble with my study. Could you give me some advice?

Mr. Liu: Yes, of course. What's the problem?

Richard: I want to learn something about programming language, but have no clue where to start.

Mr. Liu: I see. Do you want to prepare for the relevant course next term?

Richard: Yes, that's what I'm thinking about.

Mr. Liu: In that case, you can first read some books or materials about the fundamentals of programming language, such as the types of programming language, the features and applications of each type, etc. Then choose one that you are most interested in.

Richard: I got it. Thank you, Mr. Liu.

Mr. Liu: You're welcome.

3.1 Introduction to Programming Language

A program is a list of instructions or statements for directing the computer to perform a required data processing task. There are various types of programming languages that one may write for a computer. Traditionally, computer programming languages are divided into four levels, which are as follows: machine language, assembly language, high-level languages, and object-oriented programming languages.

Machine Language

Machine language is the lowest level of programming languages. Machine language is the only language that a computer can identify and carry out directly. In other words, the only programming instructions that a computer actually carries out are those written using machine language. Each machine language instruction replies a string of binary code. So we can say machine language is the language made up of binary-coded instructions that is used directly by the computer.

Each machine language instruction performs only one very low-level task. Each small step in a process must be explicitly coded in machine language. Even the small task of adding two numbers together uses three instructions written in binary, and the programmer has to remember which combination of binary digits corresponds to which instruction. So machine language is difficult to grasp and use. Further more, since machine language instruction is a string of binary code, it lacks readability, can not be easily memorized, often result in error, and the modification and debug of the programs are also very hard. In fact, very few programs are written in machine language today. Instead, most programs are written in higher-level languages and then translated into machine language, which is then executed by the computer.

Assembly Language

Assembly language appeared at the beginning of 1950s. It is the first tool developed to help the programmers. Assembly language consists of some symbols which are easy to identify and memorize. Each machine language instruction has an equivalent command in assembly language. For example, in assembly language, the statement “MOV A, B” instructs the computer to copy data from one location to another. The same instruction in machine language is a string of sixteen 0 s and 1s. The program written in assembly language is called source program of assembly language. The source program of assembly language can not be identified and carried out directly by computers. Once an assembly language source program is written, it must be firstly translated into machine language program (named object program) by another program called an assembler. Only after that, it can be carried out then. Compared with machine language, assembly language is fast and powerful. It is still difficult to use, however, because assembly language instructions are a series of abstract codes. In addition, a big drawback of assembly languages is that they are machine-oriented. Different CPUs use different machine languages and therefore there are as many assembly languages and translators as there are types of CPUs.

High-Level Languages

High-level languages were created in the middle of 1950s, which were developed to be closer to how humans think and communicate. High-level language is a kind of language that is used to compile various programs with meaningful "words" and "mathematics formulas" according to certain "grammar rules". Languages such as C, Basic, and Fortran are high-level languages.

Once a programmer has written a program in a high-level or assembly language, the program must be converted to machine code. As we have mentioned before, a program written in assembly language is input to the assembler, which translates the assembly language instructions into machine code and the machine code, which is the output from the assembler, is then executed. With high-level languages, we employ other software tools to help with the translation process. A program that translates a high-level language program into machine code is called a compiler. High-level language programs are compiled, and assembly language programs are assembled.

A program written in a high-level language can run on any computer that has an appropriate compiler for the language. Note that a compiler is a program; therefore, a machine-code version of the compiler must be available for a particular machine to be able to compile a program. Thus, to be used on multiple types of machines, each high-level language must have many compilers for that language.

High-level languages have many advantages. For example, they are easier to learn than assembly languages. They require less time to write. They provide better documentation. They are easier to maintain. A programmer skilled in writing programs in such a language is not restricted to using a single type of machine， etc.

Object-Oriented Programming Languages

Object-oriented programming languages are based on traditional high-level languages, but they enable a programmer to think in terms of collections of cooperating objects instead of lists of commands. An object is something that is seen, touched, or sensed. The types of objects may include a person, place, thing, or event. An object has attributes. Attributes are the data that represent characteristics of interest about an object. For example, the person object customer may have the following attributes: CUSTOMER NUMBER, CUSTOMER NAME, HOME ADDRESS, HOME PHONE, etc. With object-oriented programming (OOP), programmers can create relationships between one object and another. For example, objects can inherit characteristics from other objects. One of the principal advantages of object-oriented programming techniques over procedural programming techniques is that they enable programmers to create modules that do not need to be changed when a new type of object is added. A programmer can simply create a new object that inherits many of its features from existing objects. This makes object-oriented programs easier to modify. This simplifies the programmer's task, resulting in more reliable and efficient programs. To perform object-oriented programming one needs an object-oriented programming language(OOPL), such as Microsoft Visual Basic, C++, Objective C, Smalltalk, Eiffel, Common LISP Object System（CLOS）, Object Pascal, Java and Ada95, etc.

Words and Expressions

instruction [ɪn'strʌkʃ(ə)n] *n.* 指令
machine language 机器语言
assembly language 汇编语言
high-level languages 高级语言
object-oriented programming languages 面向对象的程序设计语言
identify [aɪ'dentɪfaɪ] *v.* 识别
binary ['baɪnərɪ] *adj.* 二进制的
code [kəʊd] *n.* 代码
explicit [ɪk'splɪsɪt] *adj.* 明确的，清楚的
digit ['dɪdʒɪt] *n.* 数字
readability [ˌri:də'bɪləti] *n.* 可读性
modification [ˌmɒdɪfɪ'keɪʃ(ə)n] *n.* 修改
debug [di:'bʌg] *v.* 调试
execute ['eksɪkju:t] *v.* 执行
equivalent [ɪ'kwɪv(ə)l(ə)nt] *adj.* 等价的
command [kə'mɑ:nd] *n.* 命令
compile [kəm'paɪl] *v.* 编译，编辑
source program 源程序
object program 目标程序
assembler [ə'semblə] *n.* 汇编器
abstract ['æbstrækt] *adj.* 抽象的 *n.* 抽象

drawback [drɔːbæk] *n.* 缺点，缺陷
formula ['fɔːmjʊlə] *n.* 公式
compiler [kəm'paɪlə] *n.* 编译器
attribute [ə'trɪbjuːt] *n.* 属性
characteristic [kærəktə'rɪstɪk] *n.* 特征
inherit [ɪn'herɪt] *v.* 继承
procedural [prə'siːdʒərəl] *adj.* 过程化的
module ['mɒdjuːl] *n.* 模块

Language Points

1．So we can say machine language is the language made up of binary-coded instructions that is used directly by the computer.

主句：过去分词短语 made up of binary-coded instructions…作定语，修饰 the language。定语从句 that is used directly by the computer 修饰 instructions。

译文：所以，我们可以说，机器语言是一种由二进制代码指令所组成的、由计算机直接使用的语言。

2．As we have mentioned before, **a program written in assembly language is input to the assembler**, which translates the assembly-language instructions into machine code **and the machine code**, which is the output from the assembler, **is then executed**.

主句：a program is input to the assembler and the machine code is then executed。As we have mentioned before 是非限定性定语从句，As 代后面整个主句。过去分词短语 written in assembly language 作定语，修饰 a program。which translates the assembly-language instructions into machine code 是非限定性定语从句，关系代词 which 代 the assembler。which is the output from the assembler 也是非限定性定语从句，关系代词 which 代 the machine code。

译文：正如我们前面所提到的，用汇编语言编写的程序输入汇编器，把汇编语言的指令翻译成机器代码，然后机器代码作为汇编器的输出才能执行。

3．**One of the principal advantages** of object-oriented programming techniques over procedural programming techniques **is that** they enable programmers to create modules that do not need to be changed when a new type of object is added.

主句：One of the principal advantages is that…主系表结构。of object-oriented programming techniques 介词短语作定语，修饰 advantages。over 常用作 over against，同 compared to，意思是："对照着""与……对比""不同于"，over procedural programming techniques 介词短语作状语。that 引起表语从句，表语从句中，enable sb. to do sth. 意思是"使……能够……"，从句中的主体结构是 they enable programmers to create modules。定语从句 that do not need to be changed when a new type of object is added 修饰 modules，

即关系代词 that 代 modules。when a new type of object is added 是定语从句中的时间状语从句。

译文：与面向过程的程序设计技术相比，面向对象的程序设计的主要优点是：程序员可以创建程序模块，在新对象加入时，不必修改。

3.2 C

C is relatively a minimalist programming language and it is used mostly by computer professionals to create software products. The initial development of C began in 1969. It was named "C" because many of its features were derived from an earlier language called "B". BASIC was once the leading microcomputer programming language. However, during the late 1970s, C began to take the place of BASIC and became the leading microcomputer programming language. C is a compiled language, that is, if C program runs, you must run it through a C compiler to turn the program into an executable that the computer can execute. The C program is the human-readable form, while the executable coming out of the compiler is the machine-readable and executable form. C is sometimes called "high-level language", but in fact, it is only higher-level than the various assembly languages. Compared with assembly languages, C has the following advantages. Firstly, C encourages a modular style to programming. Each chunk of code performs one and only one function. By linking these chunks of code together, programs are built. Secondly, code is usually much easier to read and write, especially for lengthy programs. Thirdly, C is much more flexible than assembly languages. Nowadays, most computer architectures are equipped with a C complier and libraries and it can be ported to any architecture. That is to say, one of the best features of C is that it is not tied to any particular hardware or system. This makes it easy for a user to write programs that will run without any changes on almost all machines. As a matter of fact, the compilers, libraries, and interpreters of higher-level languages are often implemented in C. Therefore, C is at least as portable as higher-level languages.

A newer object-oriented version of C is called C++. C++ is an enhanced version of the C language. C++ was developed in the 1980s. It is compatible with C. In fact, it is a superset of C. Therefore, existing C code can be incorporated into C++ programs. Including the basic features of C, C++ makes all C++ programs understandable to C compilers. C++ is based on the concept of object-oriented programming, which attempts to offer program data models based on real life. Therefore, C++ has features for objects, classes, and other components of an object-oriented programming (OOP). An object is defined via its class and the class determines everything about an object. Objects are individual instances of a class. For example, you may create an object "Chinese" from class "human". The Human class defines what it is to be a Human object, and all the "human-related" messages a Human object can act upon. This object has attributes of hair color, eye color and skin color, among others. The object also has ways in which the object talks, eats. C++ fully supports object-oriented programming, including the main characteristics of object-oriented development: encapsulation, data hiding, inheritance, and polymorphism. In addition, C++ also contains many improvements and features independent of object-oriented programming. There is also a visual version of the C++ language. All shows that C++ is one of the most popular programming languages for graphical applications.

The newer version of C is C#, an object-oriented programming language designed to improve productivity in the development of Web applications.

Words and Expressions

relatively ['relətɪvli]	*adv*. 相对地，比较地
minimalist ['mɪnɪm(ə)lɪst]	*n*. 极简
professional [prə'feʃ(ə)n(ə)l]	*n*. 专业人才　*adj*. 专业的，职业的
initial [ɪ'nɪʃəl]	*adj*. 最初的，开始的，初期的
compile [kəm'paɪl]	*v*. 汇集，编辑，编制；编码，编译（程序）
compiler [kəm'paɪlə]	*n*.（自动编码器）编译程序器，程序编制（器）
executable [ɪg'zekjʊtəb(ə)l]	*adj*. 可执行的，可实行的，可以作成的
modular ['mɒdjʊlə]	*adj*. 积木式的，模块化的
chunk [tʃʌŋk]	*n*. 大（厚）块；相当大的数量
flexible ['fleksɪb(ə)l]	*adj*. 灵活的，可塑造的
architecture ['ɑːkɪtektʃə]	*n*. 构造，结构
port [pɔːt]	*n*. 端口，移植
compatible [kəm'pætɪb(ə)l]	*adj*. 相容的，与……不矛盾的，一致的（with）
encapsulation [ɪnˌkæpsjʊ'leɪʃən]	*n*. 封装，封闭，密封
polymorphism [ˌpɒlɪ'mɔːfɪz(ə)m]	*n*. 多态（现象）
inheritance [ɪn'herɪt(ə)ns]	*n*. 继承
productivity [prɒdʌk'tɪvɪti]	*n*. 生产力，生产率
derive from	由……演化而来
take the place of	代替
be compatible with	与……相兼容
independent of	独立于

Language Points

1. This makes it easy for a user to write programs that will run without any changes on almost all machines.

主句：that 引导的定语从句用来修饰 programs。句中 this 指代前一句 it is not tied to any particular hardware or system（C 不与任何特定的硬件或系统绑定）。另外，此句中，it 作 make 的形式宾语，真正的宾语是不定式 to write programs。easy 作宾语补足语。it 作形式宾语时常用在此句型中。

adj.或 n.作宾语补足语，to do 作真正的宾语，it 为形式宾语，sb.作不定式的逻辑主语。句型结构为：

主语+谓语+ it + n./adj. + for sb. + to do。

译文：这使得用户编制的程序不经过任何改动就能很容易地在几乎所有计算机上运行。

2. C++ fully supports object-oriented programming, including the main characteristics of object-oriented development: encapsulation, data hiding, inheritance, and polymorphism.

译文：C++ 完全支持面向对象的程序设计，包括面向对象开发的四个特征：封装、数据隐藏、继承性和多态性。

封装（encapsulation），就是将数据与操作数据的源代码进行有机结合，形成“类”，其中数据和函数都是类的成员。封装使一部分成员充当类与外部的接口，而将其他成员隐蔽起来，以达到对成员访问权限的合理控制，使不同类之间的相互影响减少到最低限度，从而增强数据的安全性和简化程序的编写工作。

继承性（inheritance）在程序设计中是一种非常有用的特性，程序员可在既有类的基础上，只需增加或修改少量代码就能产生新的类，从而解决了代码重复使用的问题。

多态性（polymorphism）是面向对象程序设计的其中一个重要特征，指发出同样的消息被不同类型的对象接收时所产生的完全不同的行为。

3.3 Java

Java originally refers to the programming language Java developed by Sun Microsystems Inc. in May 1995 and Java platform. But here in this section, by saying Java, we mainly refer to the programming language.

Java is an object-oriented programming language which supports network computing. It originally came from “the Green Project” of Sun Microsystems Inc. in 1991. The “Green Team” led by James Gosling had 13 members. Their initial aim was to develop a distributed code system for home Consumer Electronics so that we could send e-mail to domestic appliance such as television and fridge to control them. At first, they planned to use C++, but finally dropped the plan because C++ was too complicated and was not so secure. They developed a new programming language code-named “Oak”, for there was an oak outside the window of James Gosling’s company. Later, they found a programming language called “Oak” had been existed, so one member of the team suggested naming it Java when they were drinking a kind of coffee called Java. And all the members agreed. However, “the Green Project” met some trouble—the Consumer Electronics market didn’t develop as they had expected. Luckily, with the rapid development of Internet in 1993, Sun found that Java could be used to create dynamic homepage, which gave new life to this project. In 1995, Sun announced Java publicly.

Major release versions of Java, along with their release dates are as follows:

- JDK 1.0 (January 21, 1996)
- JDK 1.1 (February 19, 1997)
- J2SE 1.2 (December 8, 1998)
- J2SE 1.3 (May 8, 2000)
- J2SE 1.4 (February 6, 2002)
- J2SE 5.0 (September 30, 2004)
- Java SE 6 (December 11, 2006)
- Java SE 7 (July 28, 2011)
- Java EE 7 (October 27, 2013)

Today, you can find Java technology in networks and devices that range from the Internet and scientific supercomputers to laptops and cell phones, from Wall Street market simulators to home game players and credit cards—just about everywhere. To be specific, it powers more than 4.5 billion devices.

Why is Java so attractive? Because it has the following features: simple, object-oriented, distributed, interpreted, robust, secure, portable, platform-independent, and multithreading.

Simple: Java was originally designed to conduct integrated control of home Consumer Electronics, so it should be simple. Its simplicity is shown on three aspects: first, it can be easily mastered by C++ programmers because of it's similarity to C++ in style; second, it abandons what can cause programming errors in C++, i.e. pointer and memory management; third, it provides various classes.

Object-oriented: Java is designed solely for object; it doesn't support procedure oriented programming like C.

Distributed: Java supports B/S computing model, thus it supports data distribution and operation distribution. Data distribution means that data can be distributed among the network hosts; while operation distribution refers to distributing a computing to different hosts.

Platform independent: Java runs on a Java Virtual Machine (JVM) which serves as an interface to different platforms. Applications written in Java can run in different software or hardware platforms without revision. It is a write once run anywhere language.

Multithreading: The multithreading of Java makes it possible to carry out many small tasks in an application at the same time. Thread, sometimes called process-lite, is a small and independent process separated from the process. A great advantage of multithreading is better interaction and real-time control.

Java is more secure compared with C++; moreover, it abandons some functions of C++, such as pointer arithmetic, construction, typedefs, etc., and requirement of freeing memory, which makes it more refined.

For more information of Java, you can log on the Java Website of Sun Microsystems Inc. at http://java.sun.com/ or http://www.javasoft.com/. For Java information in Chinese, you may find it on FTP of Tsinghua University and Chinese Academy of Sciences.

Words and Expressions

object-oriented	*adj*. 面向对象的
distributed [dɪ'strɪbjʊtɪd]	*adj*. 分布式的
code system	代码系统
Consumer Electronics	消费性电子产品
device [dɪ'vaɪs]	*n*. 装置，设置
supercomputer ['supəkəmpjutə]	*n*. 巨型（电子）计算机
simulator ['sɪmjʊleɪtə]	*n*. 模拟装置（设备）
smart card	智能卡
set-top box	置顶盒
car navigation system	汽车导航系统
lottery terminal	彩票投注中心
interpret [ɪn'tɜːprɪt]	*v*. 解释
robust [rə(ʊ)'bʌst]	*adj*. 健壮的，强壮的
portable ['pɔːtəb(ə)l]	*adj*. 可移植的
platform independent	平台无关的
multithreading	多线程的
abandon [ə'bænd(ə)n]	*v*. 放弃，丢弃
revision [rɪ'vɪʒ(ə)n]	*n*. 修改
write once run anywhere	写一次，就可在任何计算机上执行
application [ˌæplɪ'keɪʃ(ə)n]	*n*. 应用程序
Chinese Academy of Sciences	中科院

Language Points

1．Later, **they found a programming language called "Oak" had been existed**, so one member of the team suggested naming it Java when they were drinking a kind of coffee called Java.

主句：they found a programming language called "Oak" had been existed. a programming language called "Oak" had been existed 为 found 的宾语从句，called "Oak"作后置定语修饰 programming language; suggested naming it Java 建议把它(指前面的 programming language）命名为 Java，suggest 作建议，提议讲时，后面可跟 n., v-ing 或 that sb. do sth.作宾语。

译文：后来他们发现已经有一种程序设计语言叫作 Oak，所以当该团队的成员在一起喝一种叫 Java（爪哇）的咖啡时，一个组员建议把它叫 Java。

2．Luckily, with the rapid development of Internet in 1993, **Sun found that Java could be used to create dynamic homepage**, which gave new life to this project.

主句：Sun found that Java could be used to create dynamic homepage,which gave new life to this project 为非限制性定语从句，which 指代前面的整个句子 Sun found that Java could be used to create dynamic homepage。

译文：幸好，1993 网络迅速发展起来，Sun 公司发现 Java 可以用来创造动态网络主页，这给该计划带来了生机。

3．Ovum 是一家在电信、软件和 IT 咨询服务研究领域富有权威的公司，Advising on the commercial impact of technology and market changes in telecoms, software and IT services 是公司主页上的主打口号。Ovum 是总部设在欧洲的同类咨询公司中规模最大的。在软件产业方面，Ovum 的研究选取商业与技术两个视角，在战略和战术两方面展开，研究涉及 8 个主题领域：许可与定价（Licensing and Pricing）、应用生命周期（Application Lifecycle）、商业智能（Business Intelligence）、内容管理（Content Management）、基础构造（Infrastructure）、集成软件（Integration）、安全软件（Security）和 Workplace 软件。在电信领域，Ovum 更具权威性，主要从事电信与信息技术商业策略研究，Ovum 拥有 18 年协助全球电信业策略,研究、规划及国家电信法规咨询的丰富经验。Ovum 客户包括各国政府电信部门、电信运营商、通信设备制造商和其他电信产业参与者。

4．Why is Java so attractive? Because it has the following features: simple, object-oriented, distributed, interpreted, robust, secure, portable, and multithreading.

分布式（distributed）：Java 支持 B/S 计算模式，因此支持数据分布于操作分布。B/S（Browser/Server，浏览器/服务器）方式的网络结构，在客户端统一采用如 Netscape 和 IE 一类的浏览器，通过 Web 浏览器向 Web 服务器提出请求，由 Web 服务器对数据库进行操作，并将结果传回客户端。Java 提供了一个叫作 URL 的对象，利用这个对象，你可以打开并访问具有相同 URL 地址上的对象，访问方式与访问本地文件系统相同。Java 的 applet 小程序可以从服务器下载到客户端，即部分计算在客户端进行，提高系统执行效率。Java 提供了一整套网络类库，开发人员可以利用类库进行网络程序设计，方便地实现 Java 的分布式特性。

健壮的（robust）：Java 致力于检查程序在编译和运行时的错误。类型检查帮助检查出许多开发早期出现的错误。Java 自己操纵内存减少了内存出错的可能性。这些都是 Java 健壮的表现。

安全的（secure）：Java 的安全性可从两个方面得到保证：一方面，在 Java 语言里，像指针和释放内存等 C++功能被删除，避免了非法内存操作。Java 舍弃了 C++的指针对存储器地址的直接操作，程序运行时，内存由操作系统分配，这样可以避免病毒通过指针侵入系统。另一方面，当 Java 用来创建浏览器时，语言功能和一些浏览器本身提供的功能结合起来，使它更安全。Java 语言在计算机上执行前，要经过很多次的测试。它经过代码校验，检查代码段的格式，检测指针操作，对象操作是否过分以及试图改变一个对象的类型。

可移植的（portable）：同体系结构无关的特性使得 Java 应用程序可以在配备了 Java 解释器和运行环境的任何计算机系统上运行，这成为 Java 应用软件便于移植的良好基础。通过定义独立于平台的基本数据类型及其运算，Java 数据得以在任何硬件平台上保持一致。Java 语言的基本数据类型及其表示方式如下：byte 8-bit，二进制补码；short 16-bit，二进制补码；int 32-bit，二进制补码；long 64-bit，二进制补码；float 32-bit，IEEE 754 浮点数；double 32-bit，IEEE 754 浮点数；char 16-bit，Unicode 字符。在任何 Java 解释器中，数据类型都是依据以上标准具体实现的。因为几乎目前使用的所有 CPU 都能支持以上数据类型、8～64 位整数格式的补码运算和单/双精度浮点运算。

解释的（Interpreted）：Java 解释器（运行系统）能直接在任何计算机上执行目标代码指令。链接程序通常比编译程序所需资源少，所以程序员可以在创建源程序上花上更多的时间。

5．The multithreading of Java makes it possible to carry out many small tasks in an application at the same time.

主句：The multithreading of Java makes it possible to do sth. 该句中 it 作形式宾语。在英语文章中经常会出现比较长的不定式短语或从句作宾语，这时如果按正常顺序将其直接放在动词后就会造成整个句子结构失衡，因此这种情况下通常使用 it 作形式宾语，而把真正的宾语置后。此句中真正的宾语是 to carry out many small tasks in an application at the same time.

译文：Java 的多线程功能使得在一个程序里可以同时执行多个小任务。

3.4 Visual Basic

Basic was a programming language designed by two American professors at Dartmouth College in 1960s. It means Beginner's All-purpose Symbolic Instruction Code. Basic provided a friendly and non-frustrating programming environment, so it was easy to learn and convenient to use and quickly became one of the most popular programming languages. In 1980s, in order to meet the needs of structured programming, the new version of Basic language added new data types and program-controlled structures, among which True Basic, Quick Basic and Turbo Basic were better known.

In 1988, Microsoft released Windows operating system. The creation of Graphic User Interface (GUI) caused a revolution in the computer world. GUI is very popular with computer users. However, to programmers, it meant hard work to develop application software based on Windows environment. Just under this background were born visual programming languages.

Visual programming is a method of creating programs by using icons that represent common programming routines.

Visual Basic is an object-oriented programming language that was developed by Microsoft as a tool by which users of Microsoft's Windows operating system could develop their own GUI applications. It was first released in 1991 by Microsoft and the first version was similar to Quick Basic. Then Microsoft upgraded Visual Basic and it has its other versions such as Visual Basic 2.0 … Visual Basic 6.0. The latest version is VB.net. Visual Basic became very popular in the middle of 1990s.

Visual Basic has the following main features.

Visual Basic uses object-oriented programming, regards programs and data as objects and every object is visual. Programmers construct a GUI from predefined components and customize these components by describing how they should react to various events. In the case of a button, for example, the programmer would describe what should happen when that button is clicked. Therefore, programmers only write the codes of events' processes that objects should fulfill and thus raise the efficiency of programming.

Visual Basic is an event-driven programming language. Event-driven programming is very suitable for graphic user interface. Visual Basic enables the user to design the user interface quickly by drawing and arranging the user elements. Due to this, spent time is saved for the repetitive task.

Visual Basic provides easy-to-learn and easy-to-use applications development environment. It has rich data types and uses structured programming language.

Another feature of Visual Basic is Active technologies. Active technologies allow users to use the functionality provided by other applications, such as Microsoft word processor, Microsoft Excel and other Windows application. Visual Basic enables programmers to develop applications consisting of objects such as sound, graphs, animation, word processor, spreadsheet Web and so on.

Visual Basic's Internet capabilities make it easy to provide access to documents and applications across the Internet or intranet from within users' application, or to create Internet server applications.

Microsoft Visual Basic is a programming language and is considered rapid application development software that is used for creating applications. Some other popular programming languages are C++, HTML, SQL, and so on.

Words and Expressions

programming language	程序设计语言
structured programming	结构化程序设计

icon ['aɪkɒn] n. 图标
routine [ruː'tiːn] n. 常规；手续，程序
visual programming 可视化程序设计
object-oriented 面向对象的
construct [kən'strʌkt] vt. 构成；建造
customize ['kʌstəmaɪz] vt. 按客户具体要求制造
component [kəm'pəʊnənt] n. 元件，组件
repetitive [rɪ'petɪtɪv] adj. 重复性的
access ['ækses] n. 存取，访问

Language Points

1．**Visual Basic is an object-oriented programming language** that was developed by Microsoft as a tool by which users of Microsoft's Windows operating system could develop their own GUI applications.

主句：Visual Basic is an object-oriented programming language；that 引导定语从句，修饰 language；by which 也引导定语从句，修饰 tool。

译文：Visual Basic 是 Microsoft 公司开发的面向对象的程序设计语言，它是一种工具，Microsoft 公司的 Windows 操作系统用户用它来开发自己的图形用户界面应用软件。

2．In the case of a button, for example, the programmer would describe **what should happen** when that button is clicked.

主句：what should happen 是引导的名词性从句，在此作 describe 的宾语；when that button is clicked 是时间状语从句。

译文：比如一个按钮，程序员要描述当这个按钮被单击后会出现什么情况。

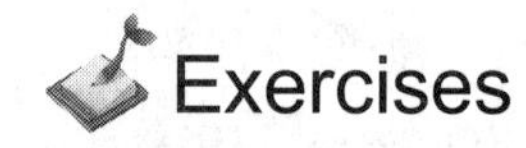

Exercises

I. Complete the following dialogue.

The problem is solved!

Richard read the materials about programming language carefully. One day he met Mr. Liu on campus, and he asked Richard how the preparation was going. Complete their dialogues by translating the Chinese into English orally.

Dialogue

Richard: Good afternoon, Mr. Liu.

Mr. Liu: Good afternoon, Richard. So have you read some materials about programming language?

Richard: Yes. I follow your advice and read ________(1)________(一些基础知识) of programming language.

Mr. Liu: For example?

Richard: For example. I know that there are four levels of programming language, _____(2)_____ (机器语言),________(3)________(汇编语言),________(4)________ (高级语言), and ________(5)________(面向对象的程序设计语言).

Mr. Liu: That's right. Anything else?

Richard: Sure! I also learnt some detailed information about the typical programming languages, such as C, Java and Visual Basic.

Mr. Liu: Really? Then can you tell me what you've got?

Richard: Of course. C is a kind of ______(6)______(高级语言). Compared with assembly language, it has such advantages as encouraging _______(7)________(模块化程序设计), code being ________(8)__________(读写通常要容易得多), especially for lengthy programs, and _______(9)________(更加灵活) than assembly languages. Java is a kind of _______(10)________(面向对象的程序设计语言) developed by Sun Microsystems Inc. It supports network computing, and was originally used to develop _______(11)________ (分布式代码系统) for home Consumer Electronics. It has a wide range of applications. The features of Java include: simple, _______(12)_______ (分布式), _______(13)________ (平台独立性), _______(14)________(多线程), etc.

Mr. Liu: What about VB? Have you read something about it?

Richard: Yes. From reading relevant materials, I know that Visual Basic is an object- oriented programming language that was developed by Microsoft. It is a tool used by Microsoft's Windows operating system users to ______(15)______(开发他们的图形用户界面应用软件). Its main features include: Visual Basic uses_____(16)______ (面向对象的程序设计方法); it is ______(17)_______(事件驱动的) programming language; it uses Active technology, etc.

Mr. Liu: Great! You really know much about programming language. The next step is to find one that you are interested in and study it further.

Richard: I will, Mr. Liu.

II. Match the English in column A with the Chinese in Column B.

A	B
instruction	模块
binary code	指令
icon	访问

source program	目标程序
object program	编译器
customize	图标
assembler	二进制代码
compiler	源程序
module	汇编器
access	自定义

III. Put the following terms into Chinese.

machine language
assembly language
high level languages
object-oriented programming languages
platform-independent
write once run anywhere
polymorphism
encapsulation
inheritance
visual programming
structured programming

IV. Translate the following sentences into Chinese.

1. So we can say machine language is the language made up of binary-coded instructions that is used directly by the computer.

2. As we have mentioned before, a program written in assembly language is input to the assembler, which translates the assembly language instructions into machine code and the machine code, which is the output from the assembler, is then executed.

3. One of the principal advantages of object-oriented programming techniques over procedural programming techniques is that they enable programmers to create modules that do not need to be changed when a new type of object is added.

4. Java is an object-oriented programming language which supports network computing.

5. Second, it abandons what can cause programming errors in C++, i.e. pointer and memory management.

6. The Human class defines what it is to be a Human object, and all the "human-related" messages a Human object can act upon.

7. In addition, C++ also contains many improvements and features independent of

object- oriented programming.

8. The newer version of C is C#, an object-oriented programming language designed to improve productivity in the development of Web applications.

9. Visual programming is a method of creating programs by using icons that represent common programming routines.

10. Visual Basic enables the user to design the user interface quickly by drawing and arranging the user elements.

V. Do you know the answers?

1. What is a program?
2. What language is the only language that a computer can identify and carry out directly?
3. What is an object? What are attributes of an object?
4. What are the characteristics of Java?
5. How does C program run?
6. What advantages does C have?
7. What features does C++ have?
8. What does Basic stand for?
9. What are the main features of Visual Basic?

Chapter 4 Database

Reading

Dialogue

What can I do?

Background: *Richard and Mark are good friends. They are in the same university. Richard is majoring in computer science and Mark is a business trade major. They often gather together to have a chat. Now let's hear what they are talking about.*

Richard: What are you up to these days?

Mark: My sister would like to open an audio-video shop. I'm helping her with it.

Richard: Selling audio-video products?

Mark: Yea, as well as rental business. And we want to manage the selling and rental business with the help of computer. You know I'm not familiar with computer well. Can you introduce something related to me?

Richard: Certainly. This task is especially suited to a database. Do you know something about database?

Mark: Little. Database, the concept seems too abstract to comprehend.

Richard: That's quite true especially for some non-computer major. I think, first of all, you have to find some materials to read.

Mark: Yes, I will. But now can you explain to me in short terms?

Richard: All right. Simply speaking, a database seems a huge storehouse of data, which is organized in a certain way waiting to be retrieved. And databases are also very common on the World Wide Web. For example, Information retrieval, E-Commerce and Dynamic Web Pages are all the examples of Web Databases in use.

Mark: It sounds interesting.

Richard: Otherwise, I suggest that you conduct business online meanwhile.

Mark: Good idea! I'll talk with my sister about it. Thank you very much!

Richard: My pleasure!

4.1 Introduction to Databases

Today we live in the information society. We use information systems to manage everything from sports statistics to payroll data. Likewise, cash registers and ATMs are supported by large information systems. They affect almost all aspects of our lives. One of the most popular general application information systems is database management systems.

A database can simply be defined as a structured set of data. A database management system (DBMS), sometimes just called a database manager, is a collection of programs that lets one or more computer users create and access data in a database.

Database systems are designed to manage large bodies of information efficiently. This purpose has led to the design of complex data structures for the representation of data in the database. Since many users of database systems are not deeply familiar with computer data structures, database developers often hide complexity through the following levels to simplify users' interactions with the system:

Physical level. The lowest level of abstraction describes how the data are actually stored. At the physical level, complex low level data structures are described in detail. The language compiler hides this level of detail from programmers.

Logical level. The next higher level of abstraction describes what data are stored in database, and what relationships exist among those data. The entire database is thus described in terms of a small number of relatively simple structures. Although implementation of the simple structures at the logical level may involve complex physical level structures, the user of the logical level does not need to be aware of this complexity. The logical level of abstraction is used by database administrators, who must decide what information is to be kept in the database. Programmers using a programming language also work at this level.

View level. The highest level of abstraction describes only part of the entire database. Despite the use of simpler structures at the logical level, some complexity remains, because of the large size of the database. Many users of the database system will not be concerned with all this information. Instead, such users need to access only a part of the database. So that their interaction with the system is simplified, the view level of abstraction is defined. The system may provide many views for the same database. In addition to hiding details of the logical level of the database, the views also provide a security mechanism to prevent users from accessing parts of the database. For example, tellers in a bank see only that part of the database that has information on customer accounts; they can not access information concerning salaries of employees.

Database systems are designed to manage large bodies of information. The management of data involves both the definition of structures for the storage of information and the provision of mechanisms for the manipulation of information. In addition, the database system must provide for the safety of the information stored, despite system crashes or attempts at unauthorized access. If data are to be shared among several users, the system must avoid possible anomalous results.

Words and Expressions

payroll ['peɪrəʊl]	*n.* 职工工资册（名单），工资单
likewise ['laɪkwaɪz]	*adv.* 同样地，照样地
database management system (DBMS)	数据库管理系统
retrieve [rɪ'triːv]	*v.* 检索
abstraction [əb'strækʃ]	*n.* 抽象
interactions [ɪntər'ækʃ(ə)nz]	*n.* 交互
language compiler	语言编译器
implementation [ɪmplɪmen'teɪʃ(ə)n]	*n.* 执行，实现
database administrator	数据库管理员
view [vjuː]	*n.* 视图
database models	数据库模型
manipulation [məˌnɪpjʊ'leɪʃ(ə)n]	*n.* 处理，操作，操纵
anomalous [ə'nɒm(ə)ləs]	*adj.* 异常的

Language Points

1. **<u>A database management system</u>** (DBMS), sometimes just called a database manager, **<u>is a collection of programs</u>** that lets one or more computer users create and access data in a database.

主句：A database management system (DBMS) is a collection of programs。sometimes just called a database manager 是过去分词短语作非限制性定语修饰 DBMS。that lets one or more computer users create and access data in a database 是定语从句修饰 programs。

译文：数据库管理系统（DBMS），有时又称数据库管理者，是一组程序，它允许一个或多个计算机用户创建数据库和访问数据库中的数据。

2. Since many users of database systems are not deeply familiar with computer data structures, **<u>database developers often hide complexity</u>** through the following levels to simplify users' interactions with the system.

主句：database developers often hide complexity。Since 引导原因状语从句。介词短语

through the following levels 作状语，表示方式、手段。不定式短语 to simplify users' interactions with the system 作状语，表目的。

译文：由于许多数据库系统的用户对于计算机的数据结构没有太深的了解，数据库开发人员经常通过如下几个层次向用户屏蔽其复杂性，以简化用户与系统之间的交互。

4.2 The Relational Database Model

A database is a structured collection of records or data. A computer database relies upon software to organize the storage of data. The software models the database structure in what are known as database models. The model in most common use today is the relational model. There are other models such as the hierarchical model, the network model and the object-oriented database model.

The terms relational, network, object-oriented, and hierarchical all refer to the way a DBMS organizes information internally. The internal organization can affect how quickly and flexibly you can extract information.

The relational model uses a collection of tables to represent both data and the relationships among those data. Each table has multiple columns, and each column has a unique name. The relational data model represents all data in the database as simple two-dimensional tables called relations. It can relate data stored in one table to data in another as long as the two tables share a common data element.

As an example, consider the database table shown in Table 4-2-1, which contains information about movies. Each row in the table corresponds to a record. Each column corresponds to a field. Each record in the table is made up of the same fields in which particular values are stored. That is, each movie record has a MovieId, a Title, a Genre, and a Rating that contain the specific data for each record. A database table is given a name, such as Movie in this case.

Table 4-2-1　Movie table

Movie			
MovieId	Title	Genre	Rating
101	Sixth Sense	Thriller Horror	PG-13
102	Back to the Future	Comedy Adventure	PG
103	Monsters, Inc.	Animation Comedy	G
104	Field of Dreams	Fantasy Drama	PG
105	Alien	Sci-fi Horror	R
106	Unbreakable	Thriller	PG-13
107	X-Men	Action Sci-fi	PG-13
5022	Elizabeth	Drama Period	R
5793	Independence Day	Action Sci-fi	PG-13
7442	Platoon	Action Drama War	R

Usually, one or more fields of a table are identified as key fields. The key field(s) uniquely identifies a record among all other records in the table, That is, the valve stored in the key field(s) for each record in a table must be unique. In the Movie table, the MovieId field would be the logical choice for a key. That way, two movies could have the same title. Certainly the Genre and Rating fields are not appropriate key fields in this case.

The movie table in Table 4-2-1, happens to be presented in the order of increasing MovieId value, but it could have been displayed in other ways, such as alphabetical by movie title. In this case, there is no inherent relationship among the rows of data in the table. Relational database tables present a logical view of the data and have nothing to do with the underlying physical organization (how the records are stored on disk).

Suppose we wanted to create a movie rental business. In addition to the list of movies for rent, we must create a database table to hold information about our customers. The Customer table in Table 4-2-2 could represent this information. Similar to what we did with our Movie table, the Customer table contains a CustomerId field to serve as a key.

Table 4-2-2 Customer table

Customer			
CustomerId	Name	Address	CreditCardNumber
101	Dennis Cook	123 Main Street	1234 5678 9876 5432
102	Doug Nickle	456 Second Ave	5678 9876 5432 1234
103	Randy Wolf	789 Elm Street	9876 5432 1234 5678
104	Amy Stevens	321 Yellow Brick Road	5432 1234 5678 9876
105	Robert Person	654 Lois Lane	1122 3344 5566 7788
106	David Coggin	987 Broadway	9988 7766 5544 3322
107	Susan Klaton	345 Easy Street	2345 6789 8765 4321

The Movie table and the Customer table show how data can be organized as records within isolated tables. The real power of relational database management systems, though, lies in the ability to create tables that conceptually link various tables together. Data from several tables are combined by a relational database management system through the fields that the tables have in common. The name relational database implies that the software relates data in different tables by common fields in those tables.

Continuing our movie rental example, we need to be able to represent the situation in which a particular customer rents a particular movie. Because “rents” is a relationship between a customer and a movie, we can represent it as a record. The date rented and the date due are attributes (the fields of a record are sometimes called the attributes of a database object) of the relationship that should be in the record. The Rents table in Table 4-2-3 contains a collection of these relationship records that represents the movies that are currently rented.

Table 4-2-3 Rents table

Rents			
CustomerId	MovieId	DateRented	DateDue
103	104	3-12-2006	3-13-2006
103	5022	3-12-2006	3-13-2006
105	107	3-12-2006	3-15-2006

The rents table contains information about the objects (a record in a database table is also called a database object or an entity) in the relationship (customers and movies), as well as the attributes of the relationship. It does not hold all of the data about a customer or a movie, however. In a relational database, we avoid duplicating data as much as possible. For instance, it is not need to store the customer's name and address in the Rent table—that data is already stored in the Customer table. The Rent table and the Customer table have the common field CustomerId. So they are related by the common field. When we need that data, we use the CustomerId stored in the Rents table to look up the customer's detailed data in the Customer table. Likewise, the Rent table and the Movie table are related by the common field MovieId. When we need data about the movie that was rented, we look it up in the Movie table using the MovieId stored in the Rents.

Note that the CustomerId value 103 is shown in two records in the Rents table. Its two appearances indicate that the same customer rented two different movies.

Data is modified in, added to, and deleted from our various database tables as needed. When movies are added or removed from the available stock, we update the records of the Movie table. As people become new customers of our store, we add them to the Customer table. On an ongoing basis， we add and remove records from the Rents table as customer rent and return videos.

Words and Expressions

relational model [rɪ'leɪʃ(ə)n(ə)l] 关系模型
hierarchical model [haɪə'rɑːkɪk(ə)l] 分层模型
network model 网状模型
object-oriented database model 面向对象的数据库模型
extract ['ekstrækt] *v.* 提取
two-dimensional 二维的
field [fiːld] *n.* 字段
genre ['ʒɒnrə] *n.*（文艺作品的）类型
rate [reɪt] *v.* 分级
identify [aɪ'dentɪfaɪ] *v.* 确定，定义

unique [juː'niːk] *adj.* 唯一的，独一无二的
alphabetical [ælfə'betɪk(ə)l] *adj.* 依字母顺序的，字母的
inherent [ɪn'hɪər(ə)nt] *adj.* 内在的
underlying physical organization 低层物理结构
irrelevant [ɪ'reləvənt] *adj.* 不相关的
isolated ['aɪsəleɪtɪd] *adj.* 独立的，单独的
attribute [ə'trɪbjuːt] *n.* 属性
object ['ɒbdʒɪkt] *n.* 对象
entity ['entɪtɪ] *n.* 实体，存在(物)
modify ['mɒdɪfaɪ] *v.* 更改，修改
update [ʌp'deɪt] *v.* 更新
ongoing ['ɒngəʊɪŋ] *adj.* 不断发展中的

4.3 Database Languages

A database system provides two different languages: one is to specify the database schema, and the other is to express database queries and updates.

Data Definition Language (DDL)

A database schema is specified by a set of definitions expressed by a special language called a data-definition language (DDL). The result of compilation of DDL statements is a set of tables that is stored in a special file called data dictionary, or data directory.

Data Manipulation Language (DML)

By data manipulation, we mean:

The retrieval of information stored in the database.

The insertion of new information into the database.

The deletion of information from the database.

The modification of information stored in the database.

A data manipulation language (DML) is a language that enables users to access or manipulate data as organized by the appropriate data model. There are basically two types:

Procedural DMLs require a user to specify what data are needed and how to get those data.

Nonprocedural DMLs require a user to specify what data are needed without specifying how to get those data.

Nonprocedural DMLs are usually easier to learn and use than procedural DMLs. However, since a user does not have to specify how to get the data, these languages may generate code that is not as efficient as that produced by procedural languages. We can remedy this difficulty through various optimization techniques.

Structured Query Language (SQL)

The Structured Query Language (SQL) is a comprehensive database language for managing relational databases. It includes statements that specify database schemas as well as statements that add, modify, and delete database content. In addition, as its name implies, SQL provides the ability to query the database to retrieve specific data.

The original version of SQL was Sequel, developed by IBM in the early 1970s. In 1986, the American National Standards Institute (ANSI) published the SQL standard, which serves as the basis for commercial database languages for accessing relational databases.

Let's look at simple queries. The *select* statement is the primary tool for this purpose. The basic select statement includes a select clause, a from clause, and a where clause:

```
select attribute-list from table-list where condition
```

The select clause determines which attributes are returned. The from clause determines what tables are used in the query. The where clause restricts the data that is returned for example:

```
select Title from Movie where Rating = 'PG'
```

The result of this query is a list of all titles from the Movie table that have a PG rating. The where clause can be eliminated if no special restrictions are necessary:

```
select Name, Address from Customer
```

This query returns the name and address of all customers in the Customer table.

SQL supports many more variations of select statements than are shown here. Likewise, SQL's insert, update and delete statements allow the data in a table to be changed.

Words and Expressions

specify ['spesɪfaɪ]	*v.* 规定，指定，确定
schema ['skiːmə]	*n.* 结构，图表
query ['kwɪərɪ]	*n.* & *v.* 查询
Data Definition Language (DDL)	数据定义语言
compilation [kɒmpɪ'leɪʃ(ə)n]	*n.* 编码，编辑
manipulation [məˌnɪpjʊ'leɪʃ(ə)n]	*n.* 处理，操作，操纵
Data Manipulation Language (DML)	数据操纵语言
retrieval [rɪ'triːvl]	*n.* 信息检索
insertion [ɪn'sɜːʃ(ə)n]	*n.* 插入
Procedural DMLs	过程化
Nonprocedural DMLs	非过程化

remedy ['remɪdɪ]　　v. 补救，纠正，改善
optimization [ˌɒptɪmaɪ'zeɪʃən]　　n. 最佳化，最优化
Structured Query Language (SQL)　　结构化查询语言
sequel ['siːkw(ə)l]　　n. 结果
restrict [rɪ'strɪkt]　　v. 限制，约束，限定
eliminate [ɪ'lɪmɪneɪt]　　v. 除去，删除

Language Points

1. **A database schema is specified by a set of definitions** expressed by a special language called a data-definition language (DDL).

主句：A database schema is specified by a set of definitions。expressed by a special language 是过去分词短语作定语，修饰 definitions。called a data-definition language (DDL) 也是过去分词短语作定语，修饰 a special language。

译文：数据库结构是由一系列定义来确定的，这些定义由称作数据定义语言(DDL)的一种特殊语言来表达。

2. **The result** of compilation of DDL statements **is a set of tables** that is stored in a special file called **data** dictionary, or data directory.

主句：The result is a set of tables。that is stored in a special file…是定语从句，修饰 tables。called data dictionary, or data directory 是过去分词短语作定语，修饰 a special file。

译文：DDL 语句的编译结果是一系列表，这些表存储在一个称作数据字典或数据目录的特殊文件中。

3. However, since a user does not have to specify how to get the data, **these languages may generate code** that is not as efficient as that produced by procedural languages.

主句：these languages may generate code…。since 引导原因状语从句。that is not as efficient as that produced by procedural languages 是定语从句，修饰 code，关系代词 that 代前面的先行词 code。that 从句中是 as…as…结构，that (the code) is not as efficient as that…，第二个普通代词 that 代 code，且后面有过去分词短语作后置定语：produced by procedural languages。

译文：然而，由于非过程化 DML 的用户不必指明如何获得数据，导致这种语言产生代码的效率不如过程化语言的高。

4.4　Database and the Web

With the growth of the Web, there has been a similar growth in services that are accessible over the Web. Many new services are Web sites that are driven from data stored in databases. Databases are extremely common on the World Wide Web. In fact, almost all companies that offer products or corporate information, online ordering, or similar activities through a Web site use a database. The most common applications involve client-server database transactions, where the user's browser is the client software.

Database applications have been around for years, and many have been deployed using network technology long before the Web existed. The point-of-service systems used by bank tellers are obvious examples of early networked database applications. Terminals are installed in bank branches, and access to the bank's central database application is provided through a wide area network. These early applications were limited to organizations that could afford the specialized terminal equipment and, in some cases, to build and own the network infrastructure.

The Web provides cheap, ubiquitous networking. It has an existing user base with standardized Web browser software that runs on a variety of ordinary computers. For developers, Web server software is freely available that can respond to requests for both documents and programs. Several scripting languages have been adapted or designed to develop programs to use with Web servers and Web protocols.

Examples of Web Databases in Use

- Information retrieval

Databases can be used for information retrieval on the Web by nature, which actually is a huge storehouse of data waiting to be retrieved. Data is stored in the database, and Web site visitors can request and view it. The information can be product information, Web pages, press releases, maps, photographs, documents, etc.

- E-Commerce and E-Business

Another widely used database application on the Web is to support and promote E-Commerce. Catalog information, pricing, customer information, shopping cart contents, and more can be stored in a database to be retrieved on demand using an appropriate script or program to link the database with the Web site.

- Dynamic Web page

Dynamic Web page is a Web page that is returned to the user with custom content (text, images, form fields, etc.) based on the results of a search or some other request that a user input. It is also known as "dynamic HTML" or "dynamic content". The "dynamic" word is used with Web sites to refer to custom results individualized to each user in contrast to the static Web pages that do not change.

How Web Databases Work

Most Web database applications bring together the Web and databases through three layers of application logic. At the base are a database management system (DBMS) and a database. At the top is the client Web browser used as an interface to the application. Between the two lies software called middleware, usually developed with a Web server-side scripting language that can interact with the DBMS, and can decode and produce HTML used for presentation in the client Web browser.

As is shown in Fig. 4-4-1, the request to retrieve information from or store data into a Web database is usually initiated by the user. Filling out a Web page form, selecting an option from a menu displayed in a Web page, or clicking an onscreen ad are common ways database requests are made. The request is received by the Web server, which then converts the request into a database query and passes it on to the database server with the help of middleware. The database server retrieves the appropriate information and returns it to the Web server (again, via middleware) where it is displayed on the user's screen as a Web page. Now the most common types of middleware used to interface between a database and a Web page are CGI and API scripts. A newer scripting language becoming increasingly more popular is PHP and ASP.

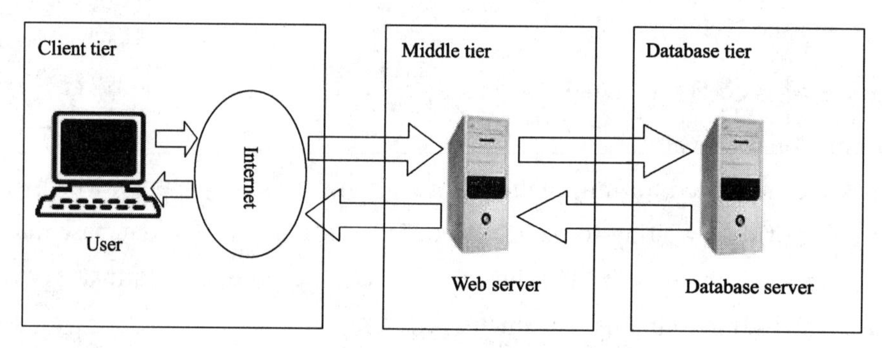

Fig. 4-4-1 How Web databases work

Words and Expressions

transaction [træn'zækʃ(ə)n]	*n.* 处理
browser ['braʊzə]	*n.* 浏览器
client ['klaɪənt]	*n.* 客户
deploy [dɪ'plɔɪ]	*v.* 展开
bank teller	银行出纳员
wide area network	广域网
infrastructure ['ɪnfrəstrʌktʃə]	*n.* 基本设施
ubiquitous [juː'bɪkwɪtəs]	*adj.* 无所不在的，普遍存在的
script [skrɪpt]	*n.* 脚本
protocol ['prəʊtəkɒl]	*n.* 协议

facilitate [fə'sɪlɪteɪt]	*v.* 促进
interactive [ˌɪntər'æktɪv]	*adj.* 交互式的
dynamic [dai'næmik]	*adj.* 动态的
press release	出版物
information retrieval	信息获取
E-commerce	电子商务
dynamic Web pages	动态网页
in contrast to	和……形成对比［对照］
static ['stætɪk]	*adj.* 静态的
layer ['leɪə]	*n.* 层
interface ['ɪntəfeɪs]	*n.* 界面，接口
interact [ɪntər'ækt]	*v.* 交互作用
decode [diː'kəʊd]	*v.* 解码
initiate [ɪ'nɪʃɪeɪt]	*v.* 开始，发起
option ['ɒpʃ(ə)n]	*n.* 选项
convert [kən'vɜːt]	*v.* 使转变，转换
middleware ['mɪdlweə]	*n.* 中间件
tier [tɪə]	*n.* 层

Language Points

1．For developers, **Web server software is freely available** that can respond to requests for both documents and programs.

主句：Web server software is freely available。freely available 的意思是随手可得。that can respond to requests for both documents and programs 是定语从句，修饰 Web server software。

译文：对于开发人员来说，可以对用户的文件和程序请求做出回答的网络服务器软件随手可得。

2．Databases **can be used for information retrieval on the Web by nature**, which actually is a huge storehouse of data waiting to be retrieved.

主句：Databases can be used for information retrieval on the Web by nature. which actually is a huge storehouse of data waiting to be retrieved 是非限定性定语从句，修饰 Databases，即关系代词 which 代 Databases。

译文：从本质上来说，数据库即可提供网络上的信息获取。实际上，它就是个等待提取的存放数据的大仓库。

3．Between <u>the two lies software</u> called middleware, usually developed with a Web server-side scripting language that can interact with the DBMS, and can decode and produce HTML used for presentation in the client Web browser.

主句：Between the two lies software。called middleware 是过去分词短语作定语，修饰 software，usually developed with a Web server-side scripting language…是过去分词短语作非限定性定语，修饰 software。that can interact with the DBMS 是定语从句，修饰 scripting language。and 并列连词，在这里连接的是两个定语从句，都修饰 scripting language，相当于：scripting language that can…, and that can…，所以 and 后面省略的主语是关系代词 that，代 scripting language。used for presentation in the client Web browser 是过去分词短语作定语，修饰 HTML。

译文：位于这两层中间的是称作中间件的软件，它通常是用网络服务器脚本语言开发的，它可以与 DBMS 交互，也可以编译并产生网页显示在客户的网络浏览器上。

Exercises

I. Complete the following dialogue.

The problem is solved!

A month later, Mark had read some materials and had some knowledge about databases. He met Richard again. Here is their talking.

Dialogue

Richard: How are things coming along?

Mark: Things are going well. I'm creating some database tables.

Richard: Oh, well, how many tables do you plan to create?

Mark: ______(1)______. (Give your own answer.)

Richard: Good. Show me your table structures.

Mark: OK. Here they are.

(How do you design the table structures? Give your own answer)__.

______________________(2)______________________.

Richard: Good job!

Mark: Thanks a lot! By the way, my sister is also interested in the online business. And we have decided to deal with some sale and rental business online in addition to the business within the shop.

Richard: Great! That means ______________(3)______________. (你们可以通过网站提供产品信息、在线订购或租赁等。)

Mark: Yes, but this needs to ______________(4)______________. (用合适的脚本语言把数据库和网站连接起来) And I know nothing about it. So I might have to ask for your help.

Richard: No problem.

II. Match the English in column A with the Chinese in Column B.

A	B
information retrieval	脚本语言
dynamic Web pages	字段
physical level	语言编译器
logical level	物理层
view level	信息获取
language compiler	视图层
record	记录
field	动态网页
browser	浏览器
scripting language	逻辑层

III. Put the following terms into Chinese.

database management system (DBMS)
database administrators
relational model
hierarchical model
network model
object-oriented database model
Data Definition Language (DDL)
Data Manipulation Language (DML)
Structured Query Language (SQL)
E-Commerce

IV. Translate the following sentences into Chinese.

1. A database management system (DBMS), sometimes just called a database manager, is a collection of programs that lets one or more computer users create and access data in a database.

2. The relational data model represents all data in the database as simple two-dimensional tables called relations. Each row in the table corresponds to a record and each column corresponds to a field.

3. A database schema is specified by a set of definitions expressed by a special language called a data-definition language (DDL). The result of compilation of DDL statements is a set of tables that is stored in a special file called data dictionary, or data directory.

4. By data manipulation, we mean:

The retrieval of information stored in the database.

The insertion of new information into the database.

The deletion of information from the database.

The modification of information stored in the database.

5. However, since a user does not have to specify how to get the data, these languages may generate code that is not as efficient as that produced by procedural languages.

6. The Structured Query Language (SQL) is a comprehensive database language for managing relational databases. It includes statements that specify database schemas as well as statements that add, modify, and delete database content. In addition, as its name implies, SQL provides the ability to query the database to retrieve specific data.

7. For developers, Web server software is freely available that can respond to requests for both documents and programs.

8. The request to retrieve information from or store data into a Web database is usually initiated by the user. Filling out a Web page form, selecting an option from a menu displayed in a Web page, or clicking an onscreen ad are common ways database requests are made. The request is received by the Web server, which then converts the request into a database query and passes it on to the database server with the help of middleware. The database server retrieves the appropriate information and returns it to the Web server (again, via middleware) where it is displayed on the user's screen as a Web page.

V. Do you know the answers?

1. How many types of languages does a database system provide?

2. What is meant by term "dynamic"?

3. What does the basic select statement include?

4. What can a database simply be defined as?

5. Must the valve stored in the key field(s) for each record in a table be unique?

6. The movie table in Table 4-2-1, happens to be presented in the order of increasing MovieId value, can it be displayed in other ways?

7. What is the difference between procedural DMLs and nonprocedural DMLs?

8. Are Databases applications common on the World Wide Web? Give some examples.

Chapter 5 Multimedia

Reading

Dialogue

What can I do?

Background: *Peter is a volunteer for the 29th Olympic Games. He takes some video about his work and some moving moments. He wants to edit the video and share his experience with his classmates after the school starts, but he doesn't know how to edit it to his requirement. So he has to ask his friend Richard for help again. He is calling Richard on the phone.*

Richard: Hello.

Peter: Hi, Richard. This is Peter. How's your holiday going?

Richard: Great!

Peter: I'm calling to ask for your help.

Richard: OK, what's that?

Peter: I want to edit the video clips captured from camera. It's about my volunteering life in Beijing Olympic Games.

Richard: I think some multimedia software would help.

Peter: Really? So you can help me, right?

Richard: Eh… To tell the truth, I'm not very familiar with the specific software. How about going to Zhongguancun Book Building tomorrow to have a check?

Peter: Good. Thank you, Richard. You are always so kind-hearted!

Richard: You are welcome. A friend in need is a friend indeed.

(Now Peter and Richard are in Zhongguancun Book Building and are reading some materials.)

5.1 Introduction to Multimedia

What Is Multimedia?

It's not easy to answer this question. So far there are several definitions of multimedia. Literally, the word "multimedia" is derived from the combination of "multi" and "media" whose core is the media. In fact, it is the combination of computer and video technology. It means that computer information can be represented through audio, graphics, images, video and animation in addition to traditional media (i.e. texts, drawings, etc.). The enabling force behind multimedia is digital technology. Multimedia represents the convergence of digital control and digital media—the PC as the digital control system and the digital media being today's most advanced form of audio and video storage and transmission. In fact, some people see multimedia simply as the marriage of PCs and video. When PC power has reached a level close to that needed for processing television and sound data streams in real time, multimedia was born.

What does Multimedia Do?

It presents information, shares ideas and emotions. It enables you to see, hear, and understand the thoughts of others. In other words, it is a form of communication. It helps people communicate ideas, marketing strategies, corporate goals, education or other information. Besides, it is also a platform for entertainment. With it, you can enjoy the beautiful songs conveniently and play interactive games.

In all, with multimedia, you don't have to be a passive recipient. You can control. You can interact. You can make it do what you want it to do. It means you can make a multimedia presentation to your own needs, and add texts, pictures, video clips or anything you like. That's the strength of multimedia and what distinguishes it from traditional media like books and television.

What does Multimedia Require?

The basic requirement of multimedia is, in fact, the standard equipment on many computers today.

First, multimedia requires sound and graphics capability. So a sound card and a graphics card are needed. Both of the sound card and graphics card are installed inside the computer's system unit. The sound card contains connectors that project from the back of the case so that you can attach speakers, headphones and a microphone. The graphics card provides a connection for the monitor's data cable.

Second, multimedia requires huge amounts of storage during both the development and application stages. Therefore, a CD-ROM drive or DVD drive are desirable. The CD serves as multimedia's chief storage and exchange medium. Without the convenient CD, many multimedia products can not be stored, and thus are unable to be distributed. A CD-ROM drive makes it possible for your computer to access audio and software CD-ROMs.

Third, multimedia requires a fast processor to handle the huge amount of data required for sound and video. With a fast processor, the computer can output smooth video sequences with a sound track that is perfectly coordinated with the action. So a speedy processor chip is necessary.

Other multimedia equipment may include capture devices such as video camera, video recorder, graphics tablets, etc.; communication networks, for example intranets and internets; and display devices such as HDTV, color printers, etc.

Words and Expressions

multimedia	多媒体
derive from	起源于，取自
combination [kɒmbɪ'neɪʃ(ə)n]	*n.* 结合，联合
video ['vɪdɪəʊ]	*n.* 视频
represent [reprɪ'zent]	*v.* 表现，描绘
audio ['ɔːdɪəʊ]	*n.* 音频，声音
graphics ['græfɪks]	*n.* 图形
image ['ɪmɪdʒ]	*n.* 图像
animation [ænɪ'meɪʃ(ə)n]	*n.* 动画
digital ['dɪdʒɪt(ə)l]	*adj.* 数字的
convergence [kən'vɜːdʒəns]	*n.* 聚合，汇聚
platform ['plætfɔːm]	*n.* 平台
interactive [ɪntər'æktɪv]	*adj.* 交互式的
recipient [rɪ'sɪpɪənt]	*n.* 接收者
clip [klɪp]	*n.* 剪辑，（影片等的）片段，摘录
graphics card	图形卡
install [ɪn'stɔl]	*v.* 安装，安置
connector [kə'nektə(r)]	*n.* 连接器，接口
drive [draɪv]	*n.* 驱动器
access ['æksɛs]	*v.* 访问，存取
processor ['prəʊsesə]	*n.* 处理器
video sequences	视频流
sound track	音轨，声轨
coordinate [kəʊ'ɔːdɪneɪt]	*v.* 使协调，使同等

chip [tʃɪp]	*n.* 芯片
capture ['kæptʃə]	*v.* 捕捉
video camera	摄像机
video recorder	录像机
graphics tablets	图形输入板
HDTV(High-Definition TV)	高清电视

Language Points

1．Literally, **the word "multimedia" is derived from the combination of "multi" and "media"** whose core is the media.

主句：the word "multimedia" is derived from the combination of "multi" and "media". 句中 multimedia 是 word 的同位语；whose core is the media 引导定语从句，修饰 multimedia。

译文：就字面意义而言，"多媒体"这个词来自"multi"和"media"的合成，其核心是媒体。

2．Multimedia **represents the convergence of digital control and digital media**—the PC as the digital control system and the digital media being today's most advanced form of audio and video storage and transmission.

主句：Multimedia represents the convergence of digital control and digital media.破折号后面的部分起解释说明作用，as 为介词，意为"作为，当作"，the digital media being today's most advanced form of audio and video storage and transmission.为独立主格结构。

译文：多媒体代表数字控制和数字媒体的汇合，个人计算机是数字控制系统，而数字媒体是当今音频和视频最先进的存储和传播形式。

3．When PC power has reached a level close to that needed for processing television and sound data streams in real time, **multimedia was born**.

主句：multimedia was born. 代词 that 代替的是 level, 过去分词短语 needed for processing television and sound data streams in real time 作后置定语，修饰 that 所代替的 level.

译文：当个人计算机的能力达到实时处理电视和声音数据流的水平时，多媒体就诞生了。

4．**That's** the strength of multimedia **and** what distinguishes it from traditional media like books and television.

主句：句型：That's…and…句中的 what distinguishes it from traditional media like books and television 从句作表语，what=the thing that，it 指代 multimedia，like 是介词，译为"例如，诸如"。

译文：这就是多媒体的力量和它与传统媒体（如书本和电视）的区别所在。

5．**A CD-ROM drive makes it possible for your computer to access** audio and software CD-ROMs.

主句：A CD-ROM drive makes it possible for your computer to do sth.句中 it 为形式宾语，真正的宾语是后面的不定式 to access audio and software CD-ROMs，your computer 是不定式中 access 的逻辑主语；该句句型结构为 v. + it + adj. + (for sb.) to do。

译文：光盘只读驱动器使计算机能够访问音频和软件光盘。

6. With a fast processor, **the computer can output smooth video sequences** with a sound track that is perfectly coordinated with the action.

主句：the computer can output smooth video sequences. 在短语 With a fast processor 中，介词 with 表示条件关系，意为“当（有）……情况下”，with a sound track 中介词 with 意为“和”；第二个 that 引导的从句为定语从句，其先行词为 sound track。

译文：安装了快速处理器，计算机就能输出更加流畅的视频流，并且保证声音和动作同步。

5.2 Uses of Multimedia

Multimedia finds its application in various areas including, but not limited to, advertisements, entertainment, art, education, engineering, medicine, mathematics, business, scientific research and spatial temporal applications. Several examples are as follows:

Multimedia in Education & Training

Learning theory in the past decade has expanded dramatically because of the introduction of multimedia. Multimedia plays a great role in school education. For teachers, it functions as a powerful auxiliary means in teaching. Compared with the traditional teaching method—chalks and blackboard, PPT courseware is more vivid and interest-inspiring. Moreover, since multimedia can present information through text, audio, graphics, images, video etc. It can help students enhance their memory. For students, multimedia has enormous potential from the point of view of self-based learning. Instead of going through the abstract information as contained in the textbook, through multimedia the learner is exposed to a variety of information which helps understand the subject more clearly. Other uses of multimedia in school include distance education, campus network, computer-based training (CBT), and so on.

Multimedia in Business

Multimedia is a powerful tool in business. Its application mainly includes training, presentation, marketing advertising, product demos, catalogues and networked communications.

CBT is also adopted by many corporations to train their employees, in-service and new.

They found it not only more lively and effective, but also economical of expenses.

Companies can create great-looking presentations through multimedia to present information to business professionals.

Compared with the traditional advertisements on television and newspapers, the multimedia-authored CD offers a new advertising form, which is more interesting and easier to be accepted. With the CDs, companies can display their products freely and thoroughly.

Fig. 5-2-1 Video conferencing

Video conferencing (Fig. 5-2-1) is a typical example of networked communications. It conducts a conference between two or more participants at different sites by using computer network to transmit audio and video data.

Multimedia in Entertainment

The field of entertainment uses multimedia extensively. When you want to listen to some music, MP3 and MP4 will be good choices; when you want to watch a movie, you can either go to the cinema or just use a cellphone; when you go to KTV with friends, you will definitely use VOD （video on demand）(Fig. 5-2-2); and when you want to play games, the importance of multimedia is even more obvious, for one of the earliest applications of multimedia was for games. With advanced multimedia technology, you can enjoy the interactive games just like you are on the scene (Fig. 5-2-3). Now, popular devices include iPhone, iPod touch(Fig. 5-2-4), iPad, etc.

Fig. 5-2-2 KTV VOD

Fig. 5-2-3 PlayStation Portable (PSP)

Fig. 5-2-4 iPod touch

Multimedia in Public Places

It is often said that this is the age of information. Getting the latest information at any place is very important for many people. Multimedia provides efficient ways to get information. No matter you are taking a bus, subway, taxi or driving your own car, you can find a large amount of information through the mobile television (Fig. 5-2-5, Fig. 5-2-6) or other multimedia equipment. Even when you are walking on the road, you can find information on the LED （large electronic display）(Fig. 5-2-7). If you are a stranger to a city, don't worry. The stand alone terminals on the road side will help you find your way. In such public places as hotels, stations, shopping malls and grocery stores, multimedia provides people with information and help at stand alone terminals or kiosks. As an effective way of communicating information, multimedia will be used in more public places.

Fig. 5-2-5 Subway mobile TV

Fig. 5-2-6 Bus mobile TV

Fig. 5-2-7 LED

Words and Expressions

auxiliary means	辅助工具
courseware	课件
enhance [ɪn'hɑ:ns]	*v.* 增强，提高
abstract ['æbstrækt]	*adj.* 抽象的
CBT (computer-based training)	基于计算机的培训
demo ['deməʊ]	*n.* 演示；示范产品
economical [i:kə'nɒmɪk(ə)l]	*adj.* 节省的，节俭的
video conferencing	视频会议
transmit [trænz'mɪt]	*v.* 传输，传播
extensive [ɪk'stensɪv]	*adj.* 大量的，范围广泛的
VOD (video on demand)	视频点播，点歌机
mobile television	移动电视
LED (large electronic display)	大型电子显示器，户外大屏幕
stand alone	孤立
terminal ['tɜ:mɪn(ə)l]	*n.* 终端
kiosk ['ki:ɒsk]	*n.* 亭子；（车站、广场等处的）广告亭，公共电话间

Language Points

1. Instead of going through the abstract information as contained in the text book,

through multimedia **the learner is exposed to a variety of information** which helps understand the subject more clearly.

主句：the learner is exposed to a variety of information. 句中的 which helps understand the subject more clearly 是修饰 information 的定语从句，help 后面的动词不定式可以带 to 也可以不带 to.

译文：利用多媒体，学生可以获得使他们更清晰理解主题的信息，而不用再冥思苦想课本上抽象的内容。

2．CBT **is also adopted by many corporations** to train their employees, in-service and new.

主句：CBT is also adopted by many corporations. 不定式 to train their employees 作目的状语，in-service and new 作定语，修饰 employees.

译文：基于计算机的培训也被很多公司采用，以培训他们的在职员工和新员工。

3．...and when you want to play games, the importance of multimedia is even more obvious, for one of the earliest applications of multimedia was for games.

主句：for one of the earliest applications...中 for 用作连词，表示“因为”。for 较 because 正式，多用于书面语中，一般都用逗号把它和前面的分句分开。

译文：当你想玩游戏时，多媒体的重要性更加明显，因为最早的多媒体程序之一就是为游戏设计的。

5.3 Multimedia Tools

To build a multimedia project, the basic tool set contains one or more authoring systems and various editing applications for text, images, sounds and videos. A few additional applications are also useful for capturing images from the screen, translating file formats, and moving files among computers for the sake of teamwork. Using these multimedia tools, your creative life will be easier.

PowerPoint

As a component of Office suite software, PowerPoint (Fig. 5-3-1) provides means for making multimedia display. Its strong functions are found in making slides. You can input title and text easily on the slide and add pictures, spreadsheets, and graphics. If you like, you can also change the layout of the slide, adjust their sequence, delete or duplicate the slide. PowerPoint is widely used for making courseware in school and preparing presentations in business. In fact, its use is far beyond that. At present, the popular versions are PowerPoint 2007，PowerPoint 2010 and PowerPoint 2016.

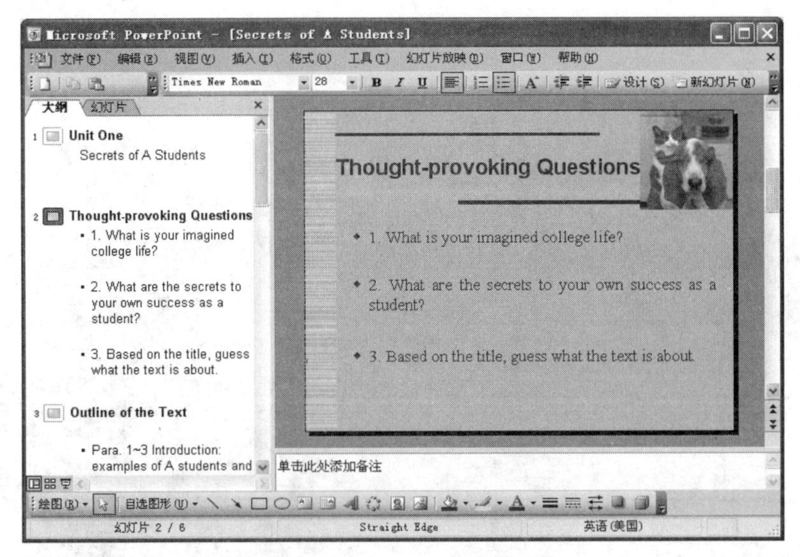

Fig. 5-3-1 PowerPoint window

Painting and Drawing Tools

Popular painting and drawing packages include CorelDraw, Illustrator, FreeHand, AutoCAD, Adobe Photoshop, Micrografx Picture Publisher, etc. They are generally divided into two categories: vector illustration programs and bitmap illustration programs. To make it easy, the difference between bitmap and vector is that bitmap images are more realistic than vector images. But bitmap images need more storage space. CorelDraw is a vector illustration program. It is popular in the fields of plane design, trademark design, freehand brushwork, package design, cartoon creation, and so on. Popular versions include CorelDraw generic version, X4 and X5, with the latest version being X6. Another professional vector graphics program is Adobe Illustrator. It is used by graphic designers, illustrators, Web designers and multimedia artists among others. However, with the development of software, the function of an illustration program will not be limited to deal with only one type of images.

3D Modeling and Animation Tools

With 3D modeling software, objects rendered in perspective appear more realistic. As long as you choose the right lightening and perspective for your final rendered image, you can create stunning scenes. Typical modeling software would be 3D Studio Max, which is often called 3ds Max or MAX for short (Fig. 5-3-2). It is a powerful, integrated 3D modeling, animation and rendering software developed by Autodesk Inc. At present, it is widely used in the fields of advertisement, film, game development, architectural design, and education,etc. Two latest versions of 3ds Max are Autodesk 3ds Max 2012 for game developers and 3ds Max Design 2012 for designers.

Some other 3D modeling, animation and rendering packages include Avid Softimage XSI,

Sumatra, Alias/Wavefront MAYA, Houdini, LightWave 3D, Animatek World Builder2C, Bryce, and Poser.

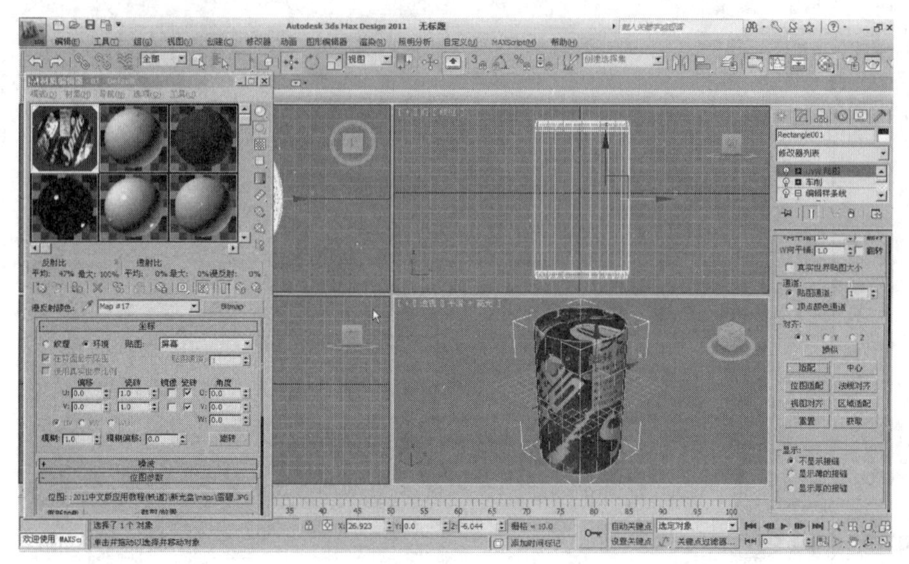

Fig. 5-3-2 3ds Max window

Video and Digital Movie Tools

To make movies from video, special movie making tools are needed. Adobe Premiere is a kind of nonlinear digital movie making software (Fig. 5-3-3). With it, you can easily edit and assemble the video clips captured from camera, other digitized movie segments, animations, scanned images, and from digitized audio or MIDI files. These functions can also be completed by MediaStudio Pro or After Effects, which are movie making tools, too. Other video and digital movie tools include Ulead Video Studio, Avid Xpress Pro HD, Sony Vegas 6.0, Digital Fusion 5.0, Discreet inferno, and so on.

Fig. 5-3-3 Adobe Premiere window

Words and Expressions

tool set	工具箱，成套工具
application [ˌæplɪ'keɪʃ(ə)n]	*n.* 应用程序，应用软件
format ['fɔːmæt]	*n.* 格式
component [kəm'pəʊnənt]	*n.* 部分，成分，元件
slide [slaɪd]	*n.* 幻灯片
spreadsheet ['spredʃiːt]	*n.* 电子数据表
layout ['leɪaʊt]	*n.* 草图，布局，版面安排
duplicate ['djuːplɪkeɪt]	*v.* 复制
vector ['vektə]	*n.* 矢量
illustration [ɪlə'streɪʃ(ə)n]	*n.* 插图，图解
bitmap ['bitmæp]	*n.* 位图
plane design	平面广告设计
freehand brushwork	写意
professional [prə'feʃ(ə)n(ə)l]	*adj.* 专业的
modeling ['mɒdəlɪŋ]	*n.* 建模
render ['rendə]	*v.* 渲染
perspective [pə'spektɪv]	*n.* 透视（画）法
stunning ['stʌnɪŋ]	*adj.* 极好的，极吸引人的
integrated ['ɪntɪgreɪtɪd]	*adj.* 完整的，综合的
nonlinear [nɒn'lɪnɪə]	*adj.* 非线性的
assemble [ə'semb(ə)l]	*v.* 组合
segment ['segm(ə)nt]	*n.* 片断

Language Points

1．To build a multimedia project, **the basic tool set contains one or more authoring systems and various editing applications** for text, images, sounds and videos.

主句：the basic tool set contains one or more authoring systems and various editing applications. 介词短语 for text, images, sounds and videos 作定语，修饰 applications。英语中，介词短语作定语都是放在被修饰的名词（或代词）后面。

译文：要创建一个多媒体项目，一套基本的工具包括一个或多个创作系统和众多的文本、 图像、声音和视频的编辑软件。

2. To make it easy, **the difference between bitmap and vector is** that bitmap images are more realistic than vector images.

主句：the difference between bitmap and vector is … 从句 that bitmap images are more realistic than vector images 为表语从句，that 引导表语从句时不可省略。

译文：简而言之，位图和矢量的区别在于位图图像比矢量图像更加真实。

3．As long as you choose the right lightening and perspective for your final rendered image, **you** can **create stunning scenes**.

主句：you can create stunning scenes. as long as 意为“只要”，引导条件状语从句；rendered 过去分词作定语，修饰 image，stunning 为现在分词作定语；当单个分词作定语时，通常放在被修饰的词前面，分词短语作定语时放在被修饰的词后面。

译文：只要为最终渲染的图像选择正确的光线和透视法，即可创造绝妙的场景。

4．With it, **you can easily edit and assemble the video clips** captured from camera, other digitized movie segments, animations, scanned images, and from digitized audio or MIDI files.

主句：you can easily edit and assemble the video clips. it 指前句中的 Adobe Premiere；captured from camera, other digitized movie segments, animations, scanned images 为分词短语作定语，修饰 video clips.

译文：利用它，能够很容易地对录像机拍摄的视频片段、其他数字化电影片段、动画、扫描图像、数字化音频或 MIDI 文件进行编辑和组合。

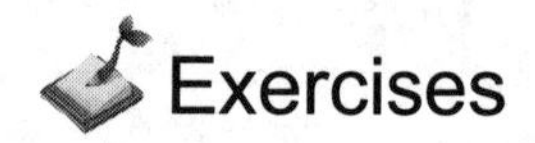

Exercises

I. Complete the following dialogue.

The problem is solved!

Peter and Richard have finished reading some materials. Now they are discussing what to buy in order to finish Peter's task. Complete their dialogues by translating the Chinese into English orally.

Dialogue

Richard: Wow, it seems multimedia has been used in almost every field of our life.

Peter: Yes. It plays a very important role in the society, especially in the fields of ______(1)______ (教育和培训), ______(2)______(商业), ______(3)______(娱乐) and public service.

Richard: So we must learn something about it.

Peter: Yes. But where should we start?

Richard: Let's start from its requirement for the computer. Here, it says the basic requirement of multimedia include______(4)______(声卡、图形卡、只读光盘驱动器和快速的处理器).

Peter: No problem with my computer.

Richard: Good. Now, here are some multimedia tools to meet your purpose.

Peter: Oh, so many multimedia packages are introduced here, for example, PowerPoint______ ______(5)______(多媒体演示软件), painting and drawing tools, ______(6)______(三维

建模和动画软件), and ____(7)____(视频和数字电影软件). Which shall we choose?

Richard: I think we need ____(8)____(视频和数字电影软件), or to be specific, Adobe Premiere. See? It is used to ____(9)____ (编辑和组合) the video clips captured from camera, other digitized movie segments, animations…

Peter: Right. So let's go to Hailong Plaza to buy it and go back home to have a try.

Richard: OK. But I think you need to buy a book on Adobe Premiere first.

II. Divide the following packages into categories.

CorelDRAW, 3ds Max, FreeHand, MediaStudio Pro, MAYA, Poser, Softimage XSI, After Effects, Illustrator, Adobe Photoshop, Adobe Premiere

Painting and Drawing Tools:________________________________

3D Modeling and Animation Tools:____________________________

Video and Digital Movie Tools:______________________________

III. Choose the most appropriate answer.

1. Multimedia equipment include（　　）.

A. a fast processor　　B. sound card and graphics card

C. CD-ROM drive　　D. all of the above

2. Which of the following statements is NOT true?（　　）

A. It's not easy to define multimedia accurately.

B. Multimedia is not only a product, but also a technology.

C. There's no relation between multimedia and the hardware.

D. Multimedia plays an important role in the modern world.

3. Many companies use（　　）to train their employees.

A. multimedia applications　　B. entertainment

C. technology　　D. animation

4.（　　）is multimedia display software.

A. Adobe Photoshop　　B. PowerPoint

C. Premiere　　D. Max

IV. Translate the following sentences into Chinese.

1. It means that computer information can be represented through audio, graphics, images, video and animation in addition to traditional media (i.e. texts, drawings, etc.).

2. The sound card contains connectors that project from the back of the case so that you can attach speakers, headphones and a microphone.

3. Multimedia requires a fast processor to handle the huge amount of data required for

sound and video.

4. Instead of going through the inane information as contained in the text book, through multimedia the learner is exposed to a variety of information which helps in a clearer understanding of the subject.

5. When you want to play games, the importance of multimedia is even more obvious, for one of the earliest applications of multimedia was for games.

6. Typical modeling software would be 3D Studio Max, which is often called 3ds Max or MAX for short.

V. Do you know the answers?

1. How many formats do you know about multimedia files?
2. Do you know some other multimedia tools?

Chapter 6 Computer Networks

Reading

Dialogue

What can I do?

Background: *Richard and Johnson are cousins. Now they are in the same university, both majoring in computer science. Johnson is a freshman, junior to Richard. One day, Richard meets Johnson. Now let's hear what they are talking about.*

Richard: Hi, Johnson, how is everything going?

Johnson: Hi, Richard, I've just adjusted to the college life.

Richard: But you look a little worried. What's up?

Johnson: I have some problems with my paper that the teacher assigned last week. I have to hand it in next Friday. There're only a few days to go, but I don't know how to start it yet. Can you lend me a hand?

Richard: Sure. What paper?

Johnson: A paper about the influence and the functions of the Internet. You know, in high school, I was just busy with study and had no time to access the Internet. I even know nothing about it.

Richard: I got it. Take it easy. It's very easy to write the paper. It's absolutely necessary for you freshmen majoring in computer science to know the influence and the functions of the Internet. Nowadays, many people take advantage of the Internet to get the information they want. In my opinion, I would recommend that in the first place, you learn to surf the Internet.

Johnson: But how?

Richard: In fact, you just key in a Web site containing a search engine where you key in the key words related to the information you want. Then, the Web pages will show you the result of browsing.

Johnson: It sounds so wonderful. I see. You are so helpful. Thank you! I really can't wait to try.

6.1 Introduction to Network

A computer network connects computers together. Through the network, users can share hardware, software, and data, as well as electronically communication with each other. It is common for computer networks to use a network server to manage the resources on a network. Network servers control access to shared printers and other hardware, as well as to shared programs and data. A server connected to the Internet to store Web pages, providing access to Web information on a network is referred to as a Web server, which particularly serves Web pages.

Types of Network

Computer networks exist in many sizes and types. A network can only be composed of a few computers, printers, and other devices. It also can consist of many small and large computers distributed over a vast geographic area. Some of the characteristics of a network include its physical arrangement, its size, and the distance it spans.

- Scale

Based on the scale or geographical range, network can be mainly classified as three types: local area networks (LAN), metropolitan area networks (MAN) and wide area networks (WAN).

Below is a further introduction of the most common types of computer networks in order of scale.

➢ Personal area network (PAN)

A personal area network (PAN) is a computer network designed for communication among computer devices close to one person. Devices employed in a PAN are printers, fax machines, telephones, scanners. A PAN typically spans within about 20–30 feet (approximately 6–9 meters). Personal area networks may be wired with computer buses. A wireless personal area network can also be made possible with network technologies such as IrDA (Infrared Data Association) and Bluetooth.

➢ Local area network (LAN)

LAN is a high-speed network covering a small geographic area, like a home, office, or building. A LAN typically connects workstations, personal computers, printers, servers, and other devices. Current LANs are most likely to be based on Ethernet technology.

➢ Campus area network (CAN)

A Campus area network (CAN) connects two or more LANs but is limited to a specific and contiguous geographical area such as a college campus or a military base.

A campus area network may be considered a type of metropolitan area network, but is generally limited to an area that is smaller than a typical metropolitan area network. This term is most often used to discuss the implementation of networks for a contiguous area. A LAN connects network devices covering a relatively short distance. A networked office building, school, or home usually contains a single LAN.

➢ Metropolitan area network (MAN)

A metropolitan area network connects two or more local area networks or campus Area Networks together, used as links between office buildings in a city but does not extend beyond the boundaries of the city.

➢ Wide area network (WAN)

A wide area network (WAN) covers a relatively broad geographic area (one city to another and one country to another country) and that often uses transmission facilities provided by common carriers, such as telephone companies. One of the most widely WANs is Internet. WAN technologies generally function at the lower three layers of the OSI reference model: the physical layer, the data link layer, and the network layer.

➢ Internetwork

Internetwork refers to any interconnection among or between public, private, commercial, industrial, or governmental networks. The interconnected networks use the Internet protocol. Depending on who administers and who participates in them, there are at least three variants of internetwork: Intranet, Extranet and Internet.

An intranet is a set of interconnected networks, using the internet protocol and uses IP-based tools such as Web browsers and ftp tools, which are under the control of a single administrative entity. That administrative entity closes the intranet to the rest of the world, and allows only specific users. Most commonly, an intranet is the private network set up by a company for its employees. A large intranet will typically have its own Web server to provide users to browse information. Intranets and extranets may or may not have connections to the Internet. If connected to the Internet, the intranet or extranet is normally protected from being accessed from the Internet without proper authorization.

An extranet is a network that is limited in scope to a single organization or entity but which also has limited connections to the networks of one or more other usually, but not necessarily, trusted organizations or entities. For example, a company's customers may be given access to some part of its intranet creating in this way an extranet; while at the same time the customers may not be considered "trusted" from a security standpoint. Technically, an extranet may also be categorized as a CAN, MAN, WAN, or other type of network, although, by definition, an extranet can not consist of a single LAN; it must have at least one connection with an external network.

➢ Internet

Internet consists of a worldwide interconnection of governmental, academic, public, and private networks based upon the Advanced Research Projects Agency Network (ARPANET) developed by DARPA of the US Department of Defense and refers to as the "Internet" with a capital "I" to distinguish it from other types of internetworks. Participants in the Internet use the Internet Protocol Suite and IP Addresses allocated by address registries. Service providers and large enterprises exchange information about the reach of their address ranges through the border gateway protocol.

- Connection method

Computer networks can also be classified according to the hardware technology that is used to connect the individual devices in the network such as optical fiber, Ethernet, wireless LAN, power line communication. Ethernet uses physical wiring to connect devices. The devices that are commonly deployed are hubs, switches, bridges, and/or routers. Wireless LAN technology is designed to connect devices without wiring. These devices use radio waves as transmission medium.

- Network topology

Computer networks can be classified according to the network topology upon which the network is based, such as bus network, star network, ring network, tree topology network, etc.

Basic Hardware Components

All networks are made up of basic hardware building blocks to interconnect network nodes, such as network interface cards（NIC）, bridges, hubs, switches, and routers.

- Network Interface Cards

A network card (Fig. 6-1-1), also called network adapter or network interface card, is a piece of computer hardware designed to allow computers to communicate over a computer network. It provides physical access to a networking medium and often provides a low-level addressing system through the use of media access control (MAC) addresses. Users can connect to each other either by using cables or wirelessly.

- Repeaters

A repeater (Fig. 6-1-2) is a device that amplifies signals along a network. It can receive a signal and retransmit it at a higher level, or onto the other side of an obstruction, so that the signal can cover longer distances without degradation. In most twisted pair Ethernet configurations, repeaters are required for cable runs longer than 100 meters.

Fig. 6-1-1 Network interface card

Fig. 6-1-2 A repeater

- Hubs

A hub (Fig. 6-1-3) contains multiple ports. When a packet arrives at one port, it is copied to all the ports of the hub for transmission. When the packets are copied, the destination address in the frame does not change to a broadcast address. It does this in a rudimentary way; it simply copies the data to all of the nodes that are connected to the hub.

- Bridges

A network bridge (Fig. 6-1-4) connects multiple network segments at the data link layer (layer 2) of the OSI model. Bridges do not promiscuously copy traffic to all ports, as hubs do, but learn which MAC (media access control) addresses are reachable through specific ports. Once the bridge associates a port and an address, it will send traffic for that address only to that port.

Fig. 6-1-3 A hub

Fig. 6-1-4 A bridge

- Switches

A switch (Fig. 6-1-5) is a device that performs switching. Specifically, it forwards and filters OSI layer 2 datagrams (chunk of data communication) between ports (connected cables) based on the MAC addresses in the packets. This is distinct from a hub in that it only forwards the datagrams to the ports involved in the communications rather than all ports connected. Strictly speaking, a switch is not capable of routing traffic based on IP address (layer 3) which is necessary for communicating between network segments or within a large or complex LAN. A switch normally has numerous ports with the intention that most or all of the networks be connected directly to another switch. Switches may operate at one or more OSI layers, including physical, data link, network, or transport. A device that operates simultaneously at more than one of these layers is called a multilayer switch.

- Routers

Routers (Fig. 6-1-6) forward data packets between networks. It can determine the best path to forward the packets. Routers work at the network layer of the TCP/IP model or layer 3 of the OSI model. Routers also provide interconnectivity between like and unlike media. Routers on a network can share information about the network. They can select the best route between any two subnets. If one part of the network is congested or out of service, a router can choose to send a packet by an alternate route. A router is connected to at least two networks, commonly two LANs or WANs or a LAN and its ISP's network. Some DSL and cable modems, for home and even office use, have been integrated with routers to allow multiple home/office computers to access the Internet through the same connection. Many of these new devices also consist of wireless access points or wireless routers to allow for IEEE 802.11b/g wireless enabled devices to connect to the network without the need for a cabled connection.

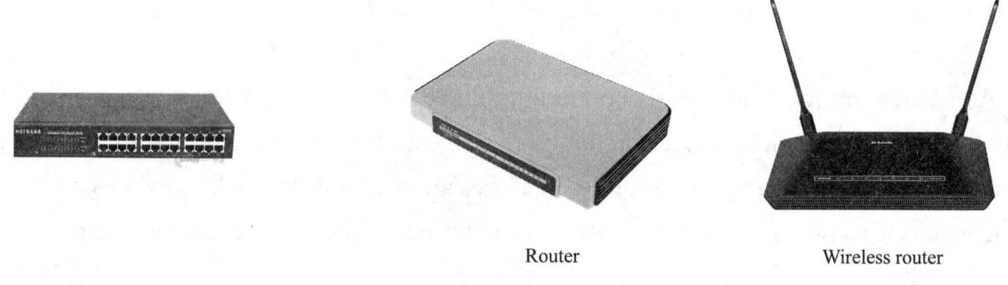

Fig. 6-1-5 A switch

Fig. 6-1-6 Routers

Protocols

A network protocol defines rules and conventions for communication between network devices. Protocols for computer networking all generally use packet switching techniques to send and receive messages in the form of packets. Network protocols include mechanisms for devices to identify and make connections with each other, as well as formatting rules that specify how data is packaged into messages sent and received. Some protocols also support message acknowledgement and data compression designed for reliable and/or high-performance network communication. Hundreds of different computer network protocols have been developed for specific purposes and environments.

Protocols exist at several levels in a telecommunication connection; the rule that layer *n* in a computer communicates with layer *n* in another computer is layer *n* protocol. Internet protocols commonly include: transmission control protocol (TCP), internet protocol (IP), and hypertext text transfer protocol (HTTP). Transmission control protocol (TCP) uses a set of rules to exchange messages with other Internet points at the information packet level; Internet text protocol (IP) uses a set of rules to send and receive massages at the Internet address level; hypertext Transfer Protocol (HTTP) is used to transmit all data present on the world wide web.

Words and Expressions

server ['sɜːvə]	*n.*服务器
distribute [dɪ'strɪbjuːt]	*v.*分配，分布，散布
geographical [dʒɪə'græfɪk(ə)l]	*adj.* 地理的，地理学的
optical ['ɒptɪk(ə)l]	*adj.* 光(学)的
metropolitan [metrə'pɒlɪt(ə)n]	*adj.* 大城市的，大都会的
medium ['miːdɪəm]	*n.* 媒介物，传导体，介质
span [spæn]	*n.* 指距，全长，跨距，一段时间，小范围 *v.* 以手指测量，跨越，架设，持续
implementation [ɪmplɪmen'teɪʃ(ə)n]	*n.* 实施，实现
contiguous [kən'tɪgjʊəs]	*adj.* 连接的，邻近的，接近的，接触的（to）
protocol ['prəʊtəkɒl]	*n.* 协议，协议列表实用程序
boundary ['baʊnd(ə)ri]	*n.* 界线，边界，境界，范围
segment ['segm(ə)nt]	*n.* 程序段
component [kəm'pəʊnənt]	*n.* 元件，组件，成分
allocate ['æləkeɪt]	*v.* 拨给，分配，配置，部署
node [nəʊd]	*n.* 节，结
interface ['ɪntəfeɪs]	*v.* 使连接，使协调
configuration [kənˌfɪgə'reɪʃ(ə)n]	*n.* 结构，构造
rudimentary [ˌruːdɪ'ment(ə)ri]	*adj.* 基本的，初步的
specify ['spesɪfaɪ]	*v.* 详细说明，具体说明
numerous ['njuːm(ə)rəs]	*adj.* [修饰单数集合名词]由多数人形成的，人数多的　[修饰复数名词]许多的，无数的
convention [kən'venʃ(ə)n]	*n.* 协定，公约，惯例
mechanism ['mek(ə)nɪz(ə)m]	*n.* 机械，机械装置[结构]
compression [kəm'preʃ(ə)n]	*n.* 压缩，压紧；浓缩；
promiscuous [prə'mɪskjʊəs]	*adj.* 杂乱的，混杂的
be integrated with	与……相一致
consist of	由……组成
a set of	一套
exchange…with…	与……交换

Language Points

1．Internet, consists of a worldwide interconnection of governmental, academic, public, and private networks based upon the Advanced Research Projects Agency Network (ARPANET) developed by DARPA of the US Department of Defense and refers to as the “Internet” with a capital “I” to distinguish it from other types of internetworks.

主句：此句看似复杂，却是一个简单句。句子的主语是 Internet，有两个并列谓语 consists of 和 refers to，其中，consists of 的宾语是一个很长的短语，a worldwide interconnection of governmental, academic, public, and private networks based upon the Advanced Research Projects Agency Network (ARPANET) developed by DARPA of the US Department of Defense，该宾语含两个过去分词短语：based upon the Advanced Research Projects Agency Network (ARPANET)和 developed by DARPA of the US Department of Defense，均作定语。

译文：互联网由全球联网的政府，学术，公共和私人网络构成，建立在美国国防部高级调查署开发的网络基础上，常用带有大写"I"的"Internet"来区别于其他种类的互联网络。

2．IrDA 是 Infrared Data Association 的缩写，它是一套使用红外线为媒介的无线传输标准。

3．Bluetooth 是一种采用 RF 射频(Radio Frequency)技术的短距离、单点多点的语音与数据信息传输交换标准。该技术的通信距离为 10 cm~10 m，如果增加信号放大装置，其通信的距离可以扩展到 100 m，并且可以绕过非金属障碍物体。

4．MAC（media access control），媒体访问控制子层协议。媒体访问控制子层协议位于 OSI 七层协议中数据链路层的下半部分，主要负责控制与连接物理层的物理介质。在发送数据的时候，MAC 协议可以事先判断是否可以发送数据，如果可以发送将给数据加上一些控制信息，最终将数据以及控制信息以规定的格式发送到物理层；在接收数据的时候，MAC 协议首先判断输入的信息是否发生传输错误，如果没有错误，则去掉控制信息发送至 LLC 层。

6.2 LAN

LAN is a kind of computer network that covers a relatively small geographic area. Local networks developed within universities&government research institutions. Most LANs are confined to a single building or group of buildings. However, LANs can be connected via telephone lines and radio waves, to exchange scientific data, and to share computing power.

LANs offer computer users many advantages, such as shared access to devices and applications, file exchange between connected users, and communication between users via E-mail and other applications. For example, a library may have a wired or wireless LAN for users to interconnect local devices such as printers and servers and to connect to the internet. On a wired LAN, PCs in the library are typically connected by cable, running the IEEE 802.3 protocol through a system of interconnection devices and eventually connect to the Internet. A wireless LAN may exist using a different IEEE protocol, 802.11b or 802.11g. The staff computers can get to the color printer, and the academic network and the Internet. Each workgroup can get to its local printer. Note that the printers are not accessible from outside their workgroup.

In contrast to wide area networks, the defining characteristics of LANs include their higher data transfer rates, smaller geographic range, and lack of a need for leased telecommunication lines. Current Ethernet or other IEEE 802.3 LAN technologies operate at speeds up to 10 Gbit/s.

- LAN devices and topologies

LAN devices often include repeaters, hubs, LAN extenders, bridges, switches, and routers. A repeater is used to interconnect the media segments of an extended network. A repeater enables a series of cable segments to be treated as a single cable. A repeater receives signals from one network segment, then amplifies, retimes and retransmits those signals to another network segment. These actions prevent signal deterioration caused by long cable lengths and large numbers of connected devices. A hub connects multiple user stations. Switches are basically multi-port bridges and work by dividing up the network into a number of segments, each of which can operate without interference from traffic local to any of the other segments. A switch learns which address resides on each of its ports and then switches data appropriately.

Each LAN has its own unique geometric arrangement of devices and LAN topologies define the manner in which network devices are organized. There are three principal topologies used in LANs: bus, ring, and star.

In bus topology (Fig. 6-2-1), all devices are connected to a central cable, called the bus or backbone. Bus networks are relatively inexpensive and easy to install for small networks. In ring topology (Fig. 6-2-2), all devices are connected to one another in the shape of a closed loop, so that each device is connected directly to two other devices, one on either side of it. Ring topologies are relatively expensive and difficult to install, but they offer high bandwidth and can span large distances. In star topology (Fig. 6-2-3), all devices are connected to a central hub. Star networks are relatively easy to install and manage, but bottlenecks can occur because all data must pass through the hub.

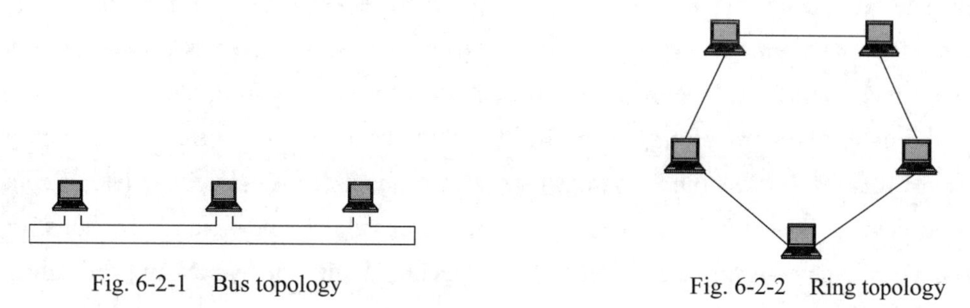

Fig. 6-2-1 Bus topology

Fig. 6-2-2 Ring topology

Fig. 6-2-4 shows the tree-topology that is commonly used in other networks.

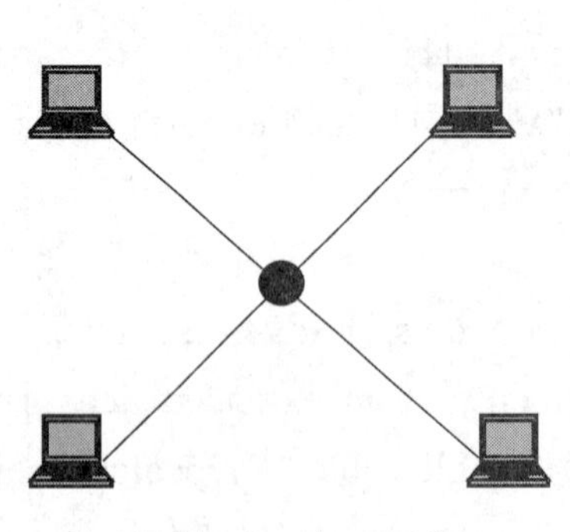

Fig. 6-2-3 Star topology

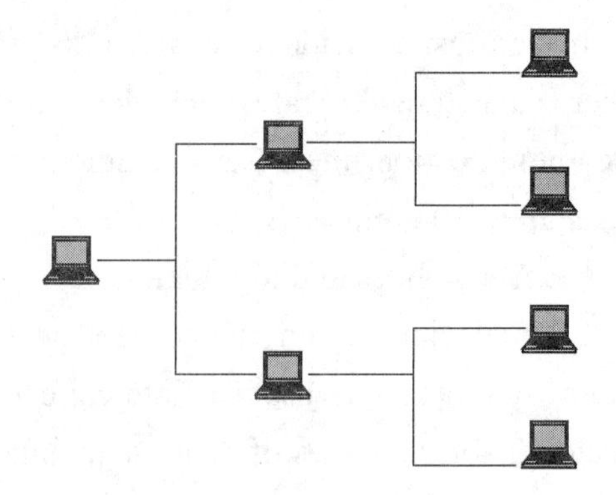

Fig. 6-2-4 Tree topology

- About ISO OSI

ISO OSI refers to International Standardization Organization/ Open System Interconnect Reference Model. It provides a complete model of the functions of a communications system. Therefore, it is one of the most important systems architectures of communications. It is a conceptual model composed of seven layers and each layer is responsible for offering certain services to the higher layers. It mainly concerns how the layers work with each other. Fig. 6-2-5 shows the seven layers of the OSI reference model.

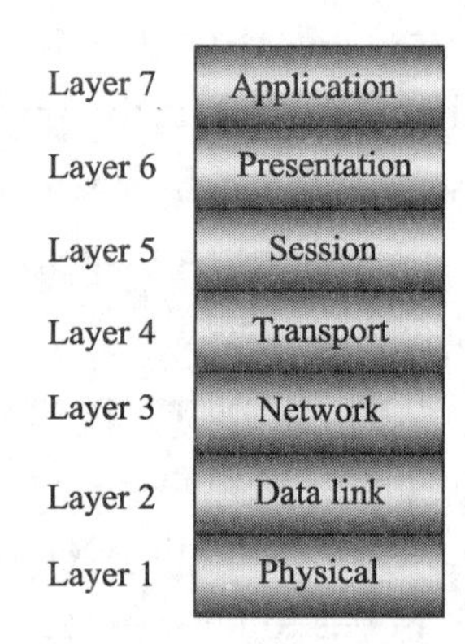

Fig. 6-2-5 Seven layers of the OSI model

In the OSI model, the seven layers can be divided into two categories: the upper layers and the lower layers. The upper layers, including application layer, presentation layer, session layer, defines the functions of the application programs. The lower four layers, the transport layer, network layer, data link layer and physical layer, deal with data transport issues. Each layer has its own set of special, related functions. The first or the lowest layer is the physical layer, which is concerned with transmitting raw bits of data over a communication channel. The second layer, called the data link layer, offers the node-to-node packet delivery, responsible for transmitting/receiving structured streams of bits over the network media. Layer 3 is the network layer. When the layer receives messages from the source host, it can convert them into packets, the units of information, and insure that all packets can be correctly received at their destinations. It is the responsibility of transport layer to guarantee successful arrival of data at the destination device. The session layer allows devices to establish and manage sessions, to allow them to exchange data. The presentation layer is concerned with the syntax and semantics of the information transmitted and is also concerned with data encryption and data security. The application layer provides services for an application program to ensure that effective communication with another application program in a network is possible.

Information being transferred from a software application in one computer system to a software application in another must pass through the OSI layers. However, layer *n* with one machine carries on a conversation with layer *n* with another machine. Exactly speaking, between computer A and B, a given message from computer A to B, there will be a flow of data through each layer at one end down through the layers in computer A and, at the other end, when the message arrives, another flow of data up through the layers in the receiving computer B and ultimately to the end user. Fig. 6-2-6 shows the conversation between the layers with one another.

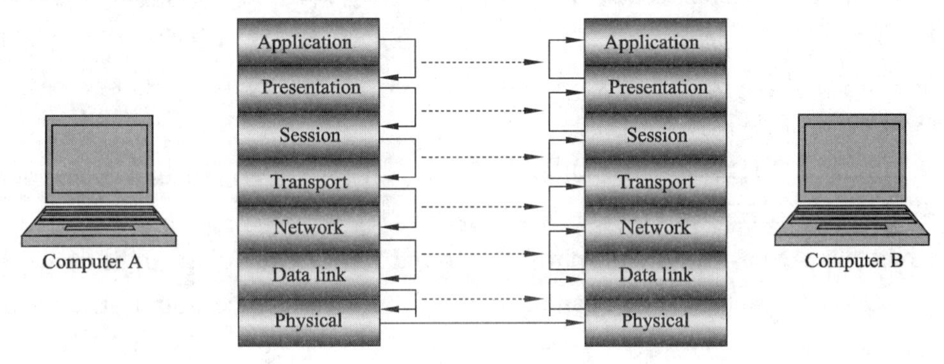

Fig. 6-2-6 Conversation between two OSI model

Words and Expressions

amplify ['æmplɪfaɪ]	*v.* 扩大，放大
deterioration [dɪˌtɪərɪə'reɪʃn]	*n.* 恶化，变坏，退化
interference [ˌɪntə'fɪər(ə)ns]	*n.* 干涉, 干预
unique [juː'niːk]	*adj.* 唯一的，独特的
geometric [ˌdʒɪə'metrɪk]	*adj.* 几何学图形的
arrangement [ə'reɪn(d)ʒm(ə)nt]	*n.* 整顿，布置，排列
principal ['prɪnsəp(ə)l]	*adj.* 主要的，最重要的，第一的
loop [luːp]	*n.* (线、带等绕成的)圈，环，线圈
bottleneck ['bɒt(ə)lnek]	*n.* 瓶颈，困难；障碍
conceptual [kən'septjʊəl]	*adj.* 概念的
layer ['leɪə]	*n.* 层
delivery [dɪ'lɪv(ə)ri]	*n.* 交付，传送
convert [kən'vɜːt]	*v.* （可）转变为
encryption [ɪn'krɪpʃən]	*n.* 编密码
bandwidth ['bændwɪtθ]	*n.* 带宽
guarantee [gær(ə)n'tiː]	*n.* 保证，担保，保证
syntax ['sɪntæks]	*n.* 句法，句子结构学
semantics [sɪ'mæntɪks]	*n.* 语义学

Language Points

1．On a wired LAN, PCs in the library are typically connected by cable, running the IEEE 802.3 protocol through a system of interconnection devices and eventually connect to the Internet.

说明：IEEE 802 是一个局域网标准系列。IEEE 是英文 Institute of Electrical and Electronics Engineers 的简称，其中文译名是电气和电子工程师协会，主要开发数据通信标准及其他标准。IEEE 802 规范定义了网卡如何访问传输介质（如光缆、双绞线、无线等），以及如何在传输介质上传输数据的方法，还定义了传输信息的网络设备之间连接建立、维护和拆除的途径。遵循 IEEE 802 标准的产品包括网卡、桥接器、路由器以及其他一些用来建立局域网络的组件。

译文：对于一个有线局域网，如图书馆的个人计算机，通常由电缆连接，通过互连的设备系统，运行 IEEE 802.3 协议，并最终连接到互联网。

2．Switches are basically multi-port bridges and work by dividing up the network into a number of segments, each of which can operate without interference from traffic local to any of the other segments.

主句：此句较长，包含了一个由 which 引导的非限制性定语从句，which 的先行词是 segments。

译文：交换机基本上是多端口网桥，工作时把网络分成很多段，每一段的本地业务运行都不会对其他网段产生干扰。

3．Information being transferred from a software application in one computer system to a software application in another must pass through the OSI layers.

主句：transfer …to…意思是“把……传送到……”。该句的谓语是 must pass。being transferred from a software application in one computer system 是过去分词短语作定语修饰 information, being transferred 用的是过去分词的现在进行时态。

译文：计算机系统上的一个软件应用向另一个系统上的应用传送出的信息必定要经过 OSI 的各个层。

6.3 Internet

Internet is a global network connecting millions of computers. Users all over the world are connected to exchange data, news and opinions to share resources and communicate with each other. The popular name for the Internet is the information superhighway. Whether you want to find the latest financial or political news, browse through library catalogs, send and receive an E-mail, chat online with colleagues, or join in a lively debate, the Internet is the tool that will take you to make it. The Internet is composed of thousands of smaller networks. Therefore, technically, no one runs the Internet. The Internet thrives and develops as its many users find new ways to create, display and retrieve the information that constitutes the Internet.

There exists no one central agency that charges individual Internet users. Rather, individuals and institutions using the Internet pay a local or regional Internet service provider for their share of services. And in turn, those smaller Internet service providers might purchase services from an even larger network. So basically, everyone who uses the Internet in some way pays for part of it. In order to be connected to Internet, users must go through service suppliers. Many options are offered with monthly rates. Depending on the option chosen, access time may vary.

- Web client/server

One of the principal uses of Internet is to share resources. This sharing is implemented by two separate programs, each running on different computers. One program, called the server, provides a particular resource. The other program, called the client, makes use of the resources. Exactly speaking, the workstation that receives network services is called a client while the computers that manage the requests for network services are called servers. If necessary to connect to another type of service， such as to set up a telnet session, or to download a file, your Web client will do this. Your Web client connects to a Web server to ask for information on your behalf.

All of the basic Internet tools, including telnet, FTP, Gopher, and the world wide web, are based on the cooperation of a client and one or more servers. In each case, users interact with the client program and it manages the details of how data is presented to users or the way in which users can look for resources. In turn, the client interacts with one or more servers where the information resides. The server receives a request, processes it, then sends a result。

- The Web / Internet

The world wide web (also be known as WWW or Web) is neither a network nor the Internet itself; it is one of the fastest-growing Internet software applications. The Web is a system of clients and servers that uses the Internet to exchange data. The Web consists of a huge collection of Web pages that available over the Internet, used to access linked documents spreading out over thousands of machines all over the Internet. Accessing the Web, people can access product information, current news, weather, airline and train schedules, various publications, music and movie downloads and so on. In addition, people can deal with types of online financial transactions, such as shopping, bank, buying and selling stock. If you want to get a page from the Web, you should type the Internet URL for the desired page. URL refers to uniform resource locator, which specifies the Internet address and filename for the Web document. If the requests from the Web browser are understood by the Web server, they should use the same standard protocol. The standard protocol for communication between a Web browser and a Web server is hypertext text transfer protocol (HTTP) that defines how the information is transferred across the network.

- IP address/domain name system

IP addresses and domain names are used to identify computers available through the Internet. IP addresses are numeric, such as 202.112.7.12. Every computer that communicates over the Internet is assigned an IP address. IP address is used to uniquely identify the device and distinguish it from other computers on the Internet.

The domain name system (DNS), consisting of its local hostname and its domain name, handles the mapping between the numerical Internet addresses and host names (domain names). For example, www.pku.edu.cn is a domain name, while 202.112.7.12 is the corresponding numerical IP address. IP address is difficult to remember while domain names are alphabetic and they're easier and more convenient to remember. A domain name includes such information as the name of the computer, and the name, type, and geographic location of the organization. The final component of a domain name, such as .com, .edu, .gov, is called the top-level domain, generally indicating the type of organization hosting the computer. For example, edu, gov, com, net, org, mil, etc. correspondingly stand for education, government, commercial, network related, organization, military, etc. It is also common to use the final two letters for indicating a country or a region. For example, uk stands for United Kindom. Currently, domain names, especially the most significant two components, for example, "sina.com", have become a valuable part of many companies' "brand". Therefore，the allocation of domain names has become highly political.

- Services provided by Internet

Currently, people can enjoy many services provided by Internet, such as FTP, E-mail, the world wide web, news, online TV/movies, online music, IRC, Telnet, etc. Many new users initially access Internet for specific Web sites and to exchange E-mail, as well as to look up information. For example, E-mail service reliably transmits and receives messages. Each message is sent from one computer to another on its way to a final destination. Online shopping is another important service. Online shopping is a fast-growing use of the Internet and it becomes more and more popular. Via Web sites, people can buy products directly from large companies or smaller retailers. Without going out of their home, they may look for whatever items they want and pay online only by clicking the mouse. When you buy what you like, you can search different shops on different websites to compare the goods and find the best price, order it and arrange to have it shipped to you overnight. The whole transaction can be done in minutes, saving a trip to the store. Shopping online can save time and money. However, online shopping has some disadvantages. The biggest disadvantage of shopping through the Internet is that you can only see pictures of a product. Therefore you will lose the pleasure of touching and trying things. You cannot check the quality of goods. Another disadvantage of shopping through the Internet is its security. There are many traps in the commercial corners of cyber space. Many people worry about the security and the reliability of the Internet. As shopping online has become more widespread, although internet users still do have security worries, it hasn't slowed down the ever-increasing numbers of online shoppers.

Words and Expressions

browse [braʊz]	*n.& v.* 浏览
catalog ['kætəlɒg]	*n.* 目录
chat [tʃæt]	*n.* 闲谈，聊天，非正式谈话
debate [dɪ'beɪt]	*n.* 讨论，争论，辩论
compose [kəm'pəʊz]	*v.* 组成，构成
thrive [θraɪv]	*v.*（throve, thrived; thriven）兴旺，繁荣，旺盛
retrieve [rɪ'triv]	*v.* 取回，找回，恢复
constitute ['kɒnstɪtjuːt]	*v.* 构成，组成
charge [tʃɑːdʒ]	*v.* 要（价），收（费）
institution [ɪnstɪ'tjuːʃ(ə)n]	*n.* 创立，设立，制定，制度，惯例，风俗
supplier [sə'plaɪə]	*n.* 供应者，补充者；
option ['ɒpʃ(ə)n]	*n.* 选择，选择权，选择自由
purchase ['pɜːtʃəs]	*v.* 购买，买
implement ['ɪmplɪm(ə)nt]	*n.* [常用复]工具；器具
mapping ['mæpɪŋ]	*n.* 映像，映射
corresponding [ˌkɒrɪ'spɒndɪŋ]	*adj.* 相当的，对应的，适合的，一致的
alphabetic [ˌælfə'betikəl]	*adj.* 照字母次序的，字母的
registrar ['redʒɪstrɑː]	*n.* 登记员
terminal ['tɜːmɪn(ə)l]	*n.* 终端，终点，极限
destination [ˌdestɪ'neɪʃ(ə)n]	*n.* 目的地，预定的目的，目标
remote [rɪ'məʊt]	*adj.* 遥远的（指时间或地点），偏僻的
on one's behalf	以某人的名义，为了某人，代表某人
carry out	执行
linked document	链接文档
distinguish ... from	把……从……当中区别出来
stand for	代表

Language Points

Whether you want to find the latest financial or political news, browse through library catalogs, send and receive an E-mail, chat online with colleagues, or join in a lively debate, the Internet is the tool that will take you to make it.

主句：此句是一个主从复合句，由 Whether 引导的状语从句 Whether you want

to find the latest financial or political news, browse through library catalogs, send and receive an e-mail, chat online with colleagues, or join in a lively debate 和主句 the Internet is the tool that will take you to make it 构成。其中，Whether 引导的状语从句中包含 want, browse, send, chat, join 并列谓语动词；主句中包含了 that 引导的定语从句。

译文：不论你要找最新的财经或政治新闻，还是浏览图书目录，或是发送、接收电子邮件，抑或与同事网上聊天，或加入到一个生动的辩论中，互联网都能帮你达到这些目的。

6.4 Network Security

With the rapid growth of the Internet and the constant introduction of new technologies, people create and enjoy new opportunities for business. On the other hand, they have to protect their computers from hackers. Intruders look for credit card numbers, bank account information, and anything else they can find through Internet. Intruders also want your computer's resources, such as the hard disk space, fast processor, and internet connection. In addition, they use these resources to attack other computers on the Internet. Therefore, network security has been a growing area of interest and become an absolutely necessary part of computer networks. Network security consists of the provisions made in an underlying computer network infrastructure, policies adopted by the network administrator to protect the network and the network-accessible resources from unauthorized access and the effectiveness of these measures combined together. Security management for networks is different for all kinds of situations. A small home or an office would only require basic security while large businesses will require high maintenance and advanced software and hardware to prevent malicious attacks from hacking and spamming. Securing your computer is not a trivial task. Some issues closely related to network security will be discussed as follows:

- Data encryption and passwords

Most security measures involved are data encryption and passwords. Data encryption is the translation of data into a form that is unintelligible without a deciphering mechanism. Only authorized people or systems can understand it. A password is a secret word or phrase that allows a user to access to a particular program or system. Each password for computers is the same as the door key. It should be unique. For each computer and service, password should be used.

- Firewall

Firewall is a collection of hardware and/or software, designed to block any unauthorized access to a private network and protect a computer or computer network from attach. System administrators often combine a proxy firewall with a packet-filtering firewall to create a highly secure system. Most home users use a software firewall. These types of firewalls can create a log file in which it records all the connection details with the PC, including connection attempts.

- Anti-virus software

Some people or companies with malicious intentions write programs like computer viruses, worms, trojan horses and spyware. These programs are all characterized as being unwanted soft wares that install themselves on users' computer through deception. Trojan horses conceal their true purpose or include a hidden functionality that a user would not want. When they are run, Trojan horses do something harmful to the computer system while they they seemingly do something useful. Worms are characterized by having the ability to replicate themselves and viruses are similar except that they achieve this by adding their code onto third party software. Once a worm has infected a computer, it would typically infect other programs and other computers. A worm might duplicate itself in a computer so often and cause the computer to crash. Viruses also slow down system performance and cause strange system behavior. In many cases viruses do serious harm to computers, either as deliberate, malicious damage or as unintentional side effects. In order to prevent damage by viruses, users typically install antivirus software, which runs in the background on the computer, detecting any suspicious software and preventing it from running.

- Honey-pot

Honey-pots, deployed in a network as surveillance and early-warning tools, mainly decoy network-accessible resources. Techniques used by the attackers who attempt to compromise these decoy resources are studied during and after an attack to keep an eye on new exploitation techniques. Such analysis could be used to further tighten security of the actual network being protected by the honey-pot.

- Anti-spyware

To install the anti-spyware software is an effective way to secure your computer. Spyware is software that runs on a computer without the explicit permission of its user. It often gathers private information from a user's computer and sends this data over the Internet back to the software manufacturer. Much like spyware, adware can run on a computer without the owner's consent. However, instead of taking information, it typically runs in the background and displays random or targeted pop-up advertisements. In many cases, this slows the computer down and may also cause software conflicts.

Nowadays, a new security technology called cloud security can identify and block threats in real time before they reach the client. It is becoming a powerful solution to protect the Internet and the clients.

Words and expressions

constant ['kɒnst(ə)nt]	*adj.* 经常的，不断的
hacker ['hækə]	*n.* 计算机窃贼，计算机新技术挑战者，黑客
intruder [ɪn'tru:də]	*n.* 闯入者，入侵者
security [sɪ'kjʊərəti]	*n.* 平安，安全（感）
infrastructure ['ɪnfrəstrʌktʃə]	*n.* 基础，防御设施
provision [prə'vɪʒ(ə)n]	*n.* 规定，条款
underlying [ʌndə'laɪɪŋ]	*adj.*（underlie 的现在分词）在下（面）的，基础的
administrator [əd'mɪnɪstreɪtə]	*n.* 管理人，治理者，行政官员
unauthorized [ˌʌn'ɔ:θəraizd]	*adj.* 未经批准[许可]的，未被授权的，越权的
effectiveness [ɪ'fektɪvnɪs]	*n.* 效率，效能，效果，有效性
maintenance ['meɪnt(ə)nəns]	*n.* 维持，保持，维修，保养
malicious [mə'lɪʃəs]	*adj.* 怀恶意的，存心不良的，有敌意的
trivial ['trɪvɪəl]	*adj.* 琐细的，不重要的
encryption [ɪn'krɪpʃən]	*n.* 编密码
unintelligible [ʌnɪn'telɪdʒɪb(ə)l]	*adj.* 无法理解的，莫名其妙的
mechanism ['mek(ə)nɪz(ə)m]	*n.* 机械，机械装置[结构]
proxy ['prɒksi]	*n.* 代理（权，人）
intention [ɪn'tenʃ(ə)n]	*n.* 意图，意向，打算，目的
install [in'stɔ:l]	*v.* 装设，装置
deception [dɪ'sepʃ(ə)n]	*n.* 瞒骗，欺诈
conceal [kən'si:l]	*v.* 隐藏，隐瞒
replicate ['replɪkeɪt]	*adj.* 复现的，复制的
deliberate [dɪ'lɪb(ə)rət]	*v.* 考虑，熟思，研究
detect [dɪ'tekt]	*v.* 发觉，发现
suspicious [sə'spɪʃəs]	*adj.* 可疑的，令人怀疑的
decoy ['di:kɒɪ]	*n.* 引诱物，饵
surveillance [sə'veɪl(ə)ns]	*n.* 监视，看守，监督，管制
compromise ['kɒmprəmaɪz]	*n.* 妥协，和解
explicit [ɪk'splɪsɪt]	*adj.* 明析的，明确的，清楚的
keep an eye on	照看，留心瞧着，注意
side effect	（药物等的）副作用
slow down	（使）慢下来

Language Points

1. Network security consists of **the provisions** made in an underlying computer network infrastructure, **policies** adopted by the network administrator to protect the network and the

network-accessible resources from unauthorized access and **the effectiveness of these measures** combined together.

主句：此句谓语consists of有三个并列宾语，即the provisions，policies，the effectiveness of these measures，并且均分别由各自的过去分词短语作定语修饰各宾语。过去分词短语made in an underlying computer network infrastructure修饰the provisions；过去分词短语adopted by the network administrator to protect the network and the network-accessible resources from unauthorized access修饰policies；过去分词短语combined together修饰the effectiveness of these measures。

译文：计算机网络防御设施里所做出的各种规定、网络管理员为保护网络和网络访问资源免遭未经授权的访问的策略以及这些措施有效的结合，便构成了网络安全。

2. Nowadays, a new security technology called cloud security can identify and block threats in real time before they reach the client. It is becoming a powerful solution to protect the Internet and the clients.

译文：目前，新兴的云安全技术可以实时确认并阻断各种威胁，不让它们到达用户计算机，成为强大的防护武器，保护着互联网和网络用户。

说明：云安全（Cloud Security）：网络时代信息安全的最新体现。它把并行处理、网格计算、未知病毒行为判断等新兴技术和概念融合在一起。通过网状的大量客户端对网络中软件行为的异常监测，以此获取互联网中木马、恶意程序的最新信息，并传送到服务器端进行自动分析和处理，再把病毒和木马的解决方案分发到每一个客户端。云安全技术的采用，使得病毒的识别和查杀不再仅仅依靠本地硬盘中的病毒库，而是依靠庞大的网络服务，实时进行采集、分析以及处理。整个互联网就是一个巨大的“杀毒软件”，参与者越多，每个参与者就越安全，整个互联网就会更安全。

Exercises

I. Complete the following dialogue.

The problem is solved!

Two days later, Richard and Johnson met again. Now let's hear what they're talking about.

Dialogue

Johnson: Hi, Richard, thanks for your help.____(1)____(访问因特网), I've got a lot of information about the influence and the functions of the Internet. Now, I get to know Internet is so important.

Richard: That's true. Nowadays, Internet is becoming increasingly significant in our lives. It has a great influence on ____(2)____（人们日常生活的许多方面）. Many people rely on Internet to know the main events____(3)____（国内外发生的）rather than newspapers or TVs.

Johnson: Yesterday, I also tried to search Internet for varied materials. I found it's a convenient and economical means of obtaining information. Instead of going to the library and checking the book one by one, ______(4)______(如你所说) the only thing I need to do is to type in the Web sites and the ______(5)______(相关关键词).

Richard: That's the case. ______(6)______(除了查询信息和发送邮件), people especially young have expanded the use of Web to other areas, such as ______(7)______(网上购物、看电视电影、网络音乐). Using Internet brings about a lot of benefits.

Johnson: That's true, Richard. But frankly speaking, it sounds a little complicated for the laymen. I'll try my best. Thank you so much, Richard.

Richard: You're welcome. Come on, Johnson.

II. Match the English in column A with the Chinese in Column B.

A	B
geographic	超文本传输协议
server	访问
access	统一资源定位器
device	服务器
URL	文件传输协议
HTTP	地理的
FTP	设备，装置

III. Put the following terms into Chinese.

LAN
MAN
WAN
Ethernet
hub
switch
bridge
router
topology
bluetooth
intranet
extranet
TCP
ISO/OSI
IP address
domain name
uploading

IV. Translate the following sentences into Chinese.

1. Computer networks can also be classified according to the hardware technology that is used to connect the individual devices in the network

2. LAN is a high-speed network covering a small geographic area, like a home, office, or building.

3. LAN is a kind of computer network that covers a relatively small geographic area.

4. LANs offer computer users many advantages, such as shared access to devices and applications, file exchange between connected users, and communication between users via E-mail and other applications.

5. These actions prevent signal deterioration caused by long cable lengths and large numbers of connected devices.

6. Star networks are relatively easy to install and manage, but bottlenecks can occur because all data must pass through the hub.

V. Do you know the answers?

1. What is network? Please briefly list the main types of networks based on the scale.

2. What is Internet?

3. Do you know what the basic hardware components that interconnect network nodes are?

4. What is the full name of HTTP?

5. What are the common Internet protocols?

6. What does ISO/OSI refer to?

7. What are the seven layers of the OSI reference model?

8. What is URL?

9. What does the basic function of the Domain Name System?

附录A 练习题参考答案

Chapter 1 Computer Hardware

Ⅰ. Complete the following dialogue.

（1）Have you got some idea about computer?

（2）buy a DIY

（3）major in

（4）assemble a computer

（5）in my opinion

（6）～（13）is omitted

Ⅱ. Look at the following illustrations and label them correctly.

1．mouse

2．printer

3．hard disk

4．monitor

5．multifunction device

6．CPU

7．memory chip

8．motherboard

9．USB flash drive

Ⅲ. Divide the following words into categories.

Input devices: keyboard, mouse, scanner, microphone

Output devices: monitor, printer, speaker, sound box

Storage units: hard disk, ROM, RAM, CD, DVD

Processor units: CPU

IV. Put the following terms into Chinese.

memory 内存

storage 外存

random access memory (RAM) 随机存储器

high resolution 高分辨率

input and output devices 输入/输出设备

liquid crystal display (LCD) 液晶显示器

absolute value 绝对值

ROM 只读存储器

CPU 中央处理器

arithmetic-logic unit 计算器

control unit 控制器

USB flash drive U 盘

V. Translate the following sentences into Chinese.

1. 计算机是快速而精准的系统，它用来接收、存储和处理数据，并在已存储的程序的指引下输出结果。

2. 当人们谈及计算机用到“内存”这个术语时，几乎总是指被称为随机存储器 RAM 的计算机主存储器。它由固定在主板上的芯片构成。

3. 在硬盘驱动器的盒子里，有一张或多张圆形的金属盘，金属盘上有许多磁道。

4. 清晰度通过分辨率显示，而分辨率又由像素决定。在显示器尺寸不变的情况下，像素越多，图像越清晰。

5. 显示器是指具有类似电视机的显像屏幕的硬件。

6. 输入设备能够将人类理解的数据和程序指令转化成计算机能处理的形式。

7. 扫描仪可以分为平板式、馈纸式、滚筒式和手持式，其中，以平板式和手持式两款最为流行。

8. 主板很重要，没有它，输入/输出设备如显示器、键盘、打印机等就不能与计算机系统交流。

9. CPU 读取并翻译软件指令，协调必须进行的处理活动。

10. RAM 存储计算机正在加工的程序和数据。

11. 执行程序的过程就是按照一定顺序处理指令的过程。

12. 声卡控制计算机的声音，显卡在屏幕上显示图像。

VI. Do you know the answers?

1. Computer hardware can be divided into four categories.

2. Through input and output devices.

3. Input devices can be classified into four types: letter input devices, pointing input devices, scanning input devices and audio-input devices.

4. Monitor and printer.

5. Combinations of input and output devices include: fax machines, multifunction devices, Internet telephony, and terminals.

6. Hard disk drive system generally provide higher speed, larger size and better reliability at a low cost than removable-medias devices. However, removable-media devices have other advantages, including the following: unlimited storage capacity, transportability, backup, security.

7. The control unit tells the rest of the computer system how to carry out a program's instructions. Under the control of this unit, programs and data are input from input devices and stored automatically and temporarily in memory, then carried out. Finally the outcome is output and printed. The arithmetic-logic unit, usually called the ALU, performs two types of operations—arithmetic and logical. Arithmetic operations perform the fundamental math operations according to arithmetic rules, such as adding, subtracting, multiplying, dividing and evaluating absolute value and so on. Logic operations are composed of comparisons.

8. Memory is an area that holds programs processed presently and data (including the results of the operation) used by programs. When the computer runs an instruction, first, it should fetch the instruction from memory, and then execute it.

9. There are three well-known types of memory chips: random-access memory (RAM), read-only memory (ROM), and complementary metal-oxide semiconductor (CMOS).

10. When people use the term "memory" in reference to computers, they are almost always referring to the computer's main memory (or primary memory) called random access memory or RAM, which is comprised of chips attached to the motherboard. Memory is sometimes referred to as temporary storage because it will be lost if the electrical power to the computer is cut off. In contrast, "storage" refers to the permanent storage available to a PC, which is also called secondary storage—usually in forms of the PC's hard drive, floppy disks and CDs. Storage is permanent, as data and programs are retained when the power is turned off.

Chapter 2 Computer Software

I. Complete the following dialogue.

（1）create and edit

（2）spelling and grammar checkers

（3）delete

（4）undelete

（5）copy and paste

（6）search and replace

（7）typeface, font size and font style

（8）format

（9）create graphs

II. Match the English in column A with the Chinese in column B.

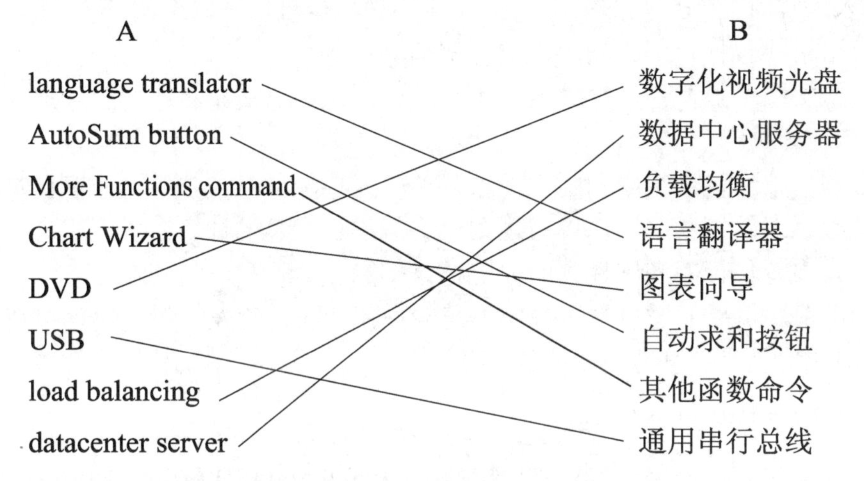

III. Put the following terms into Chinese.

系统软件

应用软件

设备驱动程序

备份软件

文件压缩软件

杀毒软件

磁盘碎片整理软件

卸载软件

文件恢复软件

图形用户界面

即插即用

命令行

源代码

换行

IV. Translate the following sentences into Chinese.

1. 计算机软件通常被分为两类：系统软件和应用软件。

2. 系统软件分为四类：操作系统、实用程序、设备驱动程序和语言翻译器。

3. 操作系统的主要功能：工作管理、内存管理和设备管理。

4．计算机系统装入一种新设备时，必须安装此设备的驱动程序，这样设备才能使用。

5．高级语言要想被计算机理解并处理，必须先翻译成机器语言。

6．图形用户界面不再是一个外壳，而被集成进了操作系统中。

7．它被广泛应用于各类计算机系统中，包括从个人计算机到大型机。

8．用文字处理软件可以生成各种类型的个人或商业信函，包括报告、通知、信件、备忘录、手稿以及其他形式的书面文件。

9．当我们输入内容时，一旦当前行输满了，换行功能会自动开始新的一行。

10．常用的段落格式包括行距、间距、对齐和缩进。

11．我们可以给文档添加背景，生成尾注和目录表，或者把其他文件的图片加入到文档中来。

12．“自动求和”按钮就是把求和函数插入某一单元格里并且建议一个求和区域。

13．利用一些数学运算符和单元格填写规则，就能进行大量的计算。

14．标题幻灯片中有加标题和副标题的位置框。

15．尽管启动屏幕还可以通过单击屏幕左下角的按钮或窗口触发，但是启动按钮是自 Windows 95 系统以来首次在任务栏上不再显示。

V. Do you know the answers?

1. A computer must be supplied with instructions so that it knows how to perform tasks. These instructions are called software. Computer software can generally be divided into two basic kinds: system software and application software.

2. System software is made up of four kinds of programs: operating system, utilities, device drivers and language translators.

3. The following are the commonly used utilities: Backup program, File compression program, Antivirus, Disk defragmenter, Uninstall program, File recovery program.

4. The three types of language translators are assembler, compiler and interpreter.

5. Windows 10 restores the Start menu that users know and love. Windows 10 start and reset quickly and help the batteries to last longer. It is designed to be compatible with all devices and Windows applications and therefore it can be used both with a keyboard or/and a touch screen.

6. Word processors allow people to create many types of personal and business communications, including reports, announcements, letters, memos, manuscripts, as well as other forms of written documents. People can create, edit, save and print them. Word processors are one of the most flexible and widely used application software tools.

7. There are five types of paragraph alignment: justified, centered, left, right and distributed.

8. Use Chart Wizard.

9. Many people employ presentation graphic software to make their report, give their lectures, present their information and so on.

10. When we want to display slides, click "View" on the menu bar. We can choose "Slide Show" or "Slide Sorter". We can use the normal view and outline view.

Chapter 3 Programming Languages

I. Complete the following dialogue.

（1）some fundamentals
（2）machine language
（3）assembly language
（4）high-level language
（5）object-oriented language
（6）high-level language
（7）a modular style to programming
（8）much easier to read and write
（9）much more flexible
（10）object-oriented programming language
（11）a distributed code system
（12）distributed
（13）platform-independent
（14）multithreading
（15）develop their own GUI applications
（16）object-oriented programming
（17）an event-driven

II. Match the English in column A with the Chinese in column B.

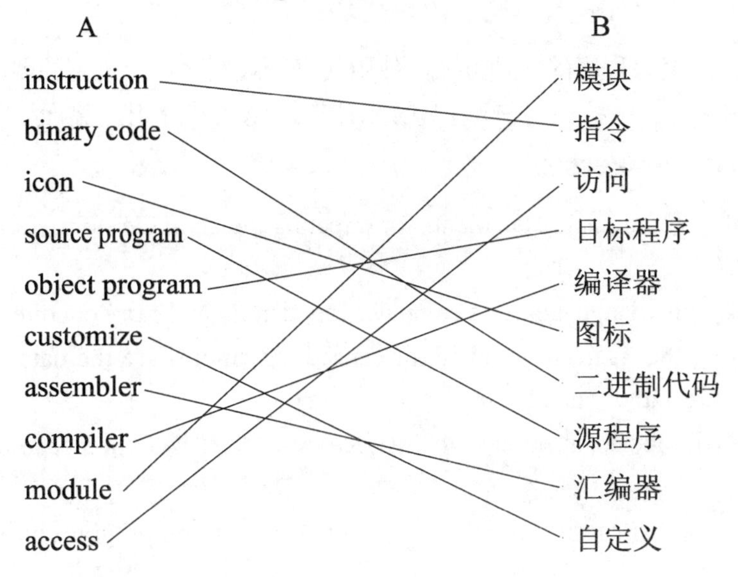

Ⅲ. Put the following terms into Chinese.

machine language　机器语言
assembly language　汇编语言
high-level languages　高级语言
object-oriented programming languages　面向对象的程序语言
platform-independent　平台无关的
write once run anywhere　写一次，即可在任何计算机上执行
polymorphism　封装
encapsulation　多态性
inheritance　继承性
visual programming　可视化程序设计
structured programming　结构化程序设计

Ⅳ. Translate the following sentences into Chinese.

1．所以，我们可以说，机器语言是一种由二进制代码指令所组成的、由计算机直接使用的语言。

2．正如我们前面所提到的，用汇编语言编写的程序输入汇编器，把汇编语言的指令翻译成机器码，然后机器码作为汇编器的输出才能执行。

3．与面向过程的程序设计技术相比，OOP 的主要优点是，程序员可以创建程序模块，在新对象加入时，不必修改。

4．Java 语言是一个支持网络计算的面向对象程序设计语言。

5．其次，Java 摒弃了 C++中容易引发程序错误的地方，如指针和内存管理。

6.“人”类定义什么是“人”对象以及“人”对象行为的所有相关信息。

7．此外，C++还拥有许多与面向对象程序设计无关的新特性和新改进。

8．C 的最新版本是 C#，是为提高开发 Web 应用软件效率而设计的一种面向对象的程序设计语言。

9．可视化程序设计语言是编写程序的一种方法，使用代表一般程序设计规则的图标。

10．Visual Basic 使得用户通过绘图和安排用户的元件来快速地设计用户界面。

Ⅴ. Do you know the answers?

1．A program is a list of instructions or statements for directing the computer to perform a required data processing task.

2．Machine language is the only language that a computer can identify and carry out directly.

3．An object is something that is seen, touched, or sensed. Attributes are the data that represent characteristics of interest about an object.

4．Simple, object-oriented, distributed, interpreted, robust, secure, portable, platform-independent and multithreading

5．If C program runs, you must run it through a C compiler to turn the program into an executable that the computer can execute.

6. C has the following advantages. Firstly, C encourages a modular style to programming. Secondly, code is usually much easier to read and write, especially for lengthy programs. Thirdly, C is much more flexible than assembly languages.

7. C++ has features for objects, classes, and other components of an object-oriented programming (OOP)

8. It stands for Beginner's All-purpose Symbolic Instruction Code.

9. Visual Basic uses object-oriented programming, regards programs and data as objects and every object is visual. It is an event-driven programming language, provides easy-to-learn and easy-to-use applications development environment, and has rich data types and uses structured programming language. It uses Active technologies. Visual Basic's Internet capabilities make it easy to provide access to documents and applications across the Internet or intranet from within users' application, or to create Internet server applications.

Chapter 4 Database

I. Complete the following dialogue.

（1），（2）略

（3）you can offer products information, online ordering or rent, etc.

（4）use appropriate scripting language to link the database with the Web site.

II. Match the English in column A with the Chinese in column B.

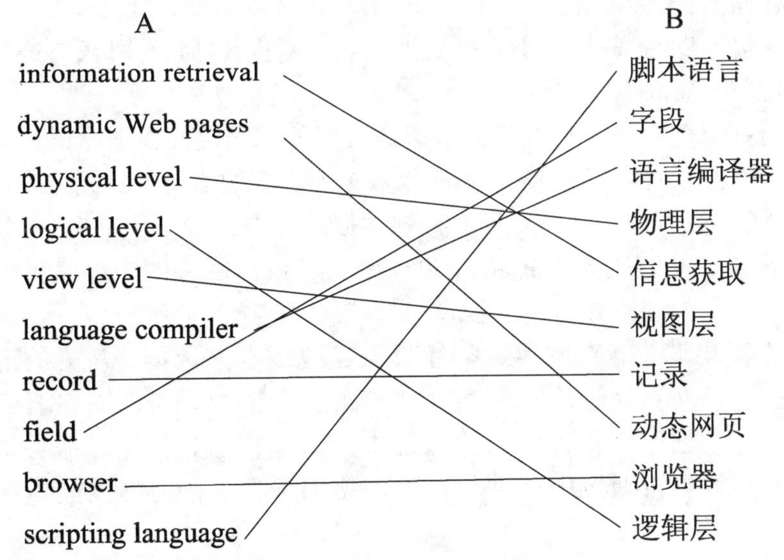

III. Put the following terms into Chinese.

database management system (DBMS) 数据库管理系统

database administrators　数据库管理员
relational model　关系模型
hierarchical model　层次模型
network model　网状模型
object-oriented database model　面对对象的数据库模型
Data Definition Language (DDL)　数据定义语言
Data Manipulation Language (DML)　数据处理语言
Structured Query Language (SQL)　结构化查询语言
E-Commerce　电子商务

IV. Translate the following sentences into Chinese.

1. 数据库管理系统（DBMS），有时称为数据库管理者，它是一组程序，让一个或多个计算机用户创建数据库和访问数据库中的数据。

2. 关系数据模型用一种称之为关系的简单的二维表来表示数据库中的所有数据。表中的每一行对应一条记录，每一列对应一个字段。

3. 数据库结构是由一系列定义来确定的，这些定义由称作数据定义语言（DDL）的一种特殊语言来表达。DDL 语句的编译结果是一系列表，这些表存储在一个称为数据字典或数据目录的特殊文件中。

4. 数据处理的意思是指：
对存储在数据库中的信息进行检索。
向数据库中插入新的信息。
从数据库中删除信息。
修改数据库中存储的信息。

5. 然而，由于非过程化 DML 的用户不必指明如何获得数据，导致这种语言产生代码的效率不如过程化语言的高。

6. 结构化查询语言（SQL）是一种管理关系型数据库的综合性的数据库语言。它包括定义数据库结构的语句，还有增加、修改和删除数据库内容的语句。除此之外，正如它的名字所暗示的，SQL 有查询数据库以提取特殊数据的能力。

7. 对于开发人员来说，可以对用户的文件和程序请求做出回答的网络服务器软件随手可得。

8. 通常由用户首先发出请求，对网络数据库提取信息或存入数据。常见的请求方式有填写网页表单，选择网页显示的菜单选项和单击屏幕上的广告。网络服务器收到请求后，将其转换成数据库查询程序，并通过中间软件传送到数据库服务器。数据库服务器提取合适的数据再通过中间件反馈给网络服务器，作为网页显示在用户显示屏上。

V. Do you know the answers?

1. A database system provides two different languages: one is to specify the database

schema, and the other is to express database queries and updates.

2. The "dynamic" word is used with Web sites to refer to custom results individualized to each user in contrast to the static Web pages that do not change.

3. The basic select statement includes a select clause, a from clause, and a where clause.

4. A database can simply be defined as a structured set of data.

5. Yes, it must.

6. Yes, it can be displayed in other ways.

7. Procedural DMLs require a user to specify what data are needed and how to get those data. Nonprocedural DMLs require a user to specify what data are needed without specifying how to get those data.

8. Yes, they are. Information retrieval, E-Commerce and Dynamic Web Pages are all the examples of Web Databases in Use.

Chapter 5 Multimedia

Ⅰ. Complete the following dialogue.

（1）education&training

（2）business

（3）entertainment

（4）sound card, graphics card, CD-ROM drive and a speedy processor

（5）a multimedia display software

（6）3D modeling and animation tools

（7）video and digital movie tools

（8）video and digital movie tools

（9）edit and assemble

Ⅱ. Divide the following packages into categories.

Painting and Drawing Tools: CorelDRAW, FreeHand, Illustrator, Adobe Photoshop.

3D Modeling and Animation Tools: 3ds Max, MAYA, Poser, Softimage XSI.

Video and Digital Movie Tools: MediaStudio Pro, After Effects, Adobe Premiere.

Ⅲ. Choose the most appropriate answer.

1. D　2. C　3. A　4. B

Ⅳ. Translate the following sentences into Chinese.

1. 这就是意味着除了通过传统媒介（如文本、图画等）外，计算机信息可以通过音频、图形、图像、视频和动画等方式呈现出来。

2. 声卡在机箱背面提供外部接口，可以把扬声器、耳机和麦克风等插到该接口。

3．多媒体需要快速的处理器来处理声音和视频所需的大量数据。

4．利用多媒体，学生无须再为教材上空洞的信息发愁，他们可以找到有助于更清晰理解主题的内容。

5．当你想玩游戏时，多媒体的重要性更加明显，因为最早的多媒体程序之一就是为游戏设计的。

6．典型的建模软件是 3D Studio Max，通常简称为 3ds Max 或 MAX。

V．Do you know the answers?

1．多媒体文件类型众多，大致分音频、图像和视频等几类，声音文件最基本的格式是 WAV（波形）格式，此外还有 MP3, MP4, MID 等格式；图像文件的基本格式是 BMP 位图格式，视频格式有 AVI 格式和 RMVB 等。还有一些压缩格式，如 JPG, GIF, TGA 等。

2．一些其他的多媒体制作软件（包）还有 Flash used specifically for Web designing, Authorware, Canvas, Cinebench, Adobe Director, Multimedia Builder 等。

Chapter 6　Computer Networks

I．Complete the following dialogue.

（1）Surfing the Internet

（2）many aspects of people's daily life.

（3）happening home and abroad

（4）as you say

（5）relevant keywords

（6）In addition to looking up information and sending E-mails

（7）online shopping, online TV and movies and online music

II．Match the English in column A with the Chinese in column B.

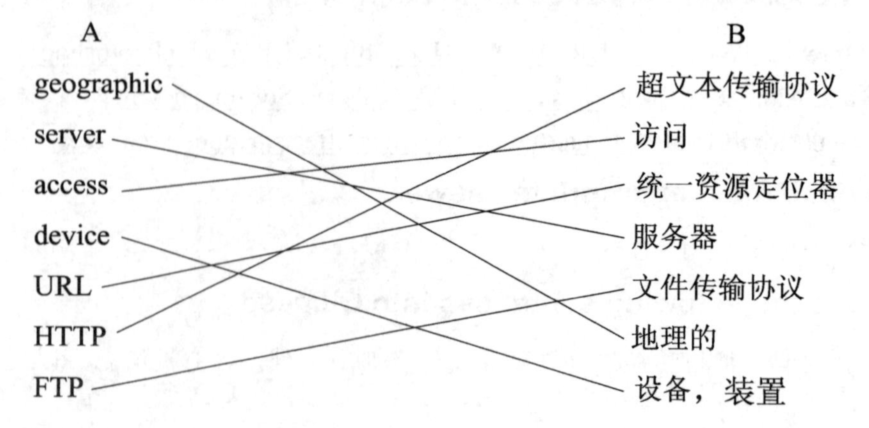

Ⅲ. Put the following terms into Chinese.

LAN	局域网
MAN	城域网
WAN	广域网
Ethernet	以太网
hub	集线器
switch	网络交换机
bridge	网桥
router	路由器
topology	拓扑技术
bluetooth	蓝牙
intranet	内联网
extranet	外联网
TCP	传输控制协议
ISO/OSI	国际标准化组织/开放系统互连参考模型
IP address	互联网协议地址
domain name	域名
uploading	上传

Ⅳ. Translate the following sentences into Chinese.

1．计算机网络也可以根据用以连接网络中单个设备的硬件技术来归类。

2．局域网是一种高速网络，涵盖一个较小的地理区域，例如一个家、一间办公室或一栋大楼。

3．局域网是指覆盖在相关小范围内的计算机网络。

4．局域网给计算机用户提供许多优势，如共享访问设备和应用软件、连网的用户之间交换文件、通过电子邮件和其他软件在用户间通信。

5．通过这种方式避免信号在长距离电缆和大量连接设备中减弱。

6．星状拓扑结构相对来说易于安装和管理，但由于所有数据都必须通过中心的集线器, 容易造成堵塞。

Ⅴ. Do you know the answers?

1．A computer network connects computers together. Through the network, users can share hardware, software and data, as well as electronically communication with each other. Based on the scale or geographical range, network can be classified as three types: Personal Area Network (PAN), Local Area Networks (LAN), Campus Area Network (CAN), Metropolitan Area Networks (MAN) and Wide Area Networks (WAN).

2．A specific internetwork, consisting of a worldwide interconnection of governmental, academic, public, and private networks based upon the Advanced Research Projects Agency

Network (ARPANET) developed by DARPA of the US Department of Defense and referred to as the “Internet” with a capital “I” to distinguish it from other types of internetworks.

3. Network interface cards, bridges, hubs, switches and routers are the basic hardware components that interconnect network nodes.

4. The full name of HTTP is Hypertext Transfer Protocol.

5. Internet protocols commonly include: Transmission Control Protocol (TCP), Internet Protocol (IP), Hypertext Transfer Protocol (HTTP).

6. ISO/OSI refers to International Standardization Organization/Open System Interconnect Reference Model.

7. The seven layers include application layer, presentation layer, session layer, the transport layer, network layer, data link layer and physical layer.

8. URL refers to Uniform Resource Locator, which specifies the Internet address and filename for the WWW document.

9. The Domain Name System handles the mapping between the numerical Internet addresses and host names (domain names).

第1章　计算机硬件

计算机是快速而精准的系统，它用来接收、存储和处理数据，并在已存储的程序的指引下输出结果。

个人计算机系统包括两个基本部分——硬件和软件。硬件是系统的物理部件，是看得见摸得着的；软件是指控制硬件操作的程序。

计算机硬件可以分为四个部分：中央处理器、存储设备、输入/输出设备。

计算机系统的基本结构图如 1-1-1 所示。

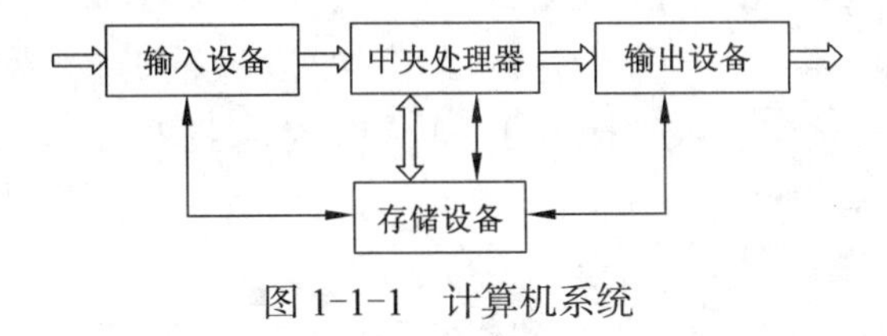

图 1-1-1　计算机系统

1.1　输入设备和输出设备

如何将数据存入 CPU？又如何将信息取出呢？本节就来介绍一下能实现计算机和人类交流的设备。输入设备能够将人类理解的数据和程序指令转化成计算机能处理的形式。输出设备恰恰相反，可将计算机处理过的信息转化成人类能够理解的形式。

输入设备

将新信息或输入的数据存入计算机的方法很多。根据所起的具体作用不同，输入设备可以分成四类：字符输入设备、指向输入设备、图像和视频输入设备以及语音输入设备。

典型的字符输入设备是键盘。它也是最常用的输入设备之一。键盘上具有能够输入字符（字母、数字和标点符号）和特殊命令的键。按键盘上的相应键能指示计算机做什么或写什么。一般而言，传统的101键位键盘有四个键区：位于键盘顶部的功能键区、打字键区、包含上下左右四个箭头移动光标的光标/编辑键区和数字键区（图 1-1-2）。新型的 104 键位键盘增加了三个Windows 菜单快捷键（图 1-1-3）。除了传统类型的键盘以外，还有折叠键盘（图 1-1-4）、人体工程学键盘（图 1-1-5）。常见的键盘品牌有罗技、微软、飞利浦、优派、双飞燕等。

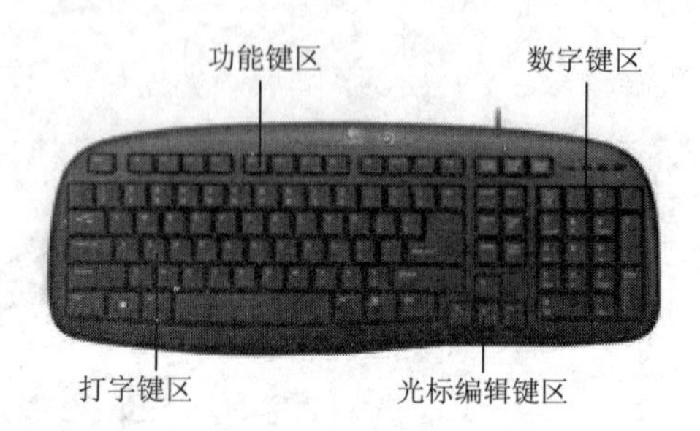

图 1-1-2　101 键位键盘

图 1-1-3　104 键位键盘

图 1-1-4　折叠键盘

图 1-1-5　人体工程学键盘

最常用的指向输入设备是鼠标。鼠标控制显示器上的指针，该指针通常以箭头的形式出现。鼠标上通常有一个、两个或三个按键，这些按键用来选择要执行的命令和控制显示在显示器上的信息。鼠标大致可以分为三类：第一类，机械鼠标（图 1-1-6）。它的底部有一个可以滚动的小球，在平滑表面滚动小球，显示器上代表鼠标的指针则会随之移动。第二类，光电鼠标（图 1-1-7）。它也是目前应用最广的鼠标。它通过发出和感应光来判断鼠标的移动，可以应用于任何表面，且这种鼠标的精度较高。第三类，无线鼠标（图 1-1-8）。它通过无线电波或红外线与计算机交流。常用的鼠标品牌与键盘相似。

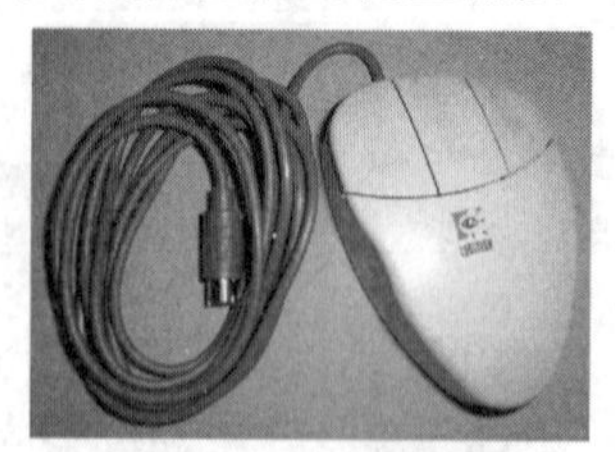

图 1-1-6　机械鼠标

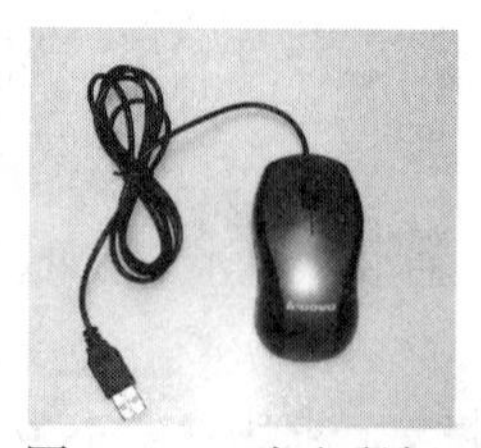

图 1-1-7　光电鼠标

图 1-1-8　无线鼠标

其他指向输入设备包括轨迹球、触摸屏、定位杆、操纵杆、光笔、触摸板和 3D 鼠标。

图像和视频输入设备可以将外界的视频或图像转化为数字化信息存入计算机。这些信息根据用户的具体要求可以保存为多种形式，常见的种类包括：数码照相机、摄像机、网络摄像头、图像扫描器、指纹扫描器、3D 扫描器、条形码阅读器和激光测距仪。用户通过扫描仪可以将书面文档、图片和其他图像输入计算机系统。扫描仪可以分为平板式、馈纸式、滚筒式和手持式（图 1-1-9），其中，以平板式和手持式两款最为流行。大部分扫描仪都有光学识别系统（OCR），该系统能够使计算机识别所扫描的内容。光学识别最典型的应用就是零售店里记录顾客购买情况的条形码阅读器（图 1-1-10）。扫描仪的著名品牌有惠普、爱普生、力捷、佳能等。

语音输入设备主要是指声音和音乐输入设备。声音输入通常由话筒和合适的软件完成，如 IBM 语音识别系统 ViaVoice 或语音识别 Dragon Naturally Speaking。音乐可以通过 MP3 或 MP4 播放器输入个人计算机，而 MP3 或 MP4 播放器同时又是输出设备。

平板式扫描仪

馈纸式扫描仪

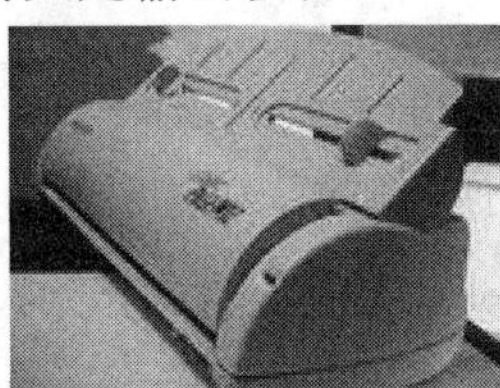

滚筒式扫描仪

手持式扫描描仪

图 1-1-9　扫描仪的类型

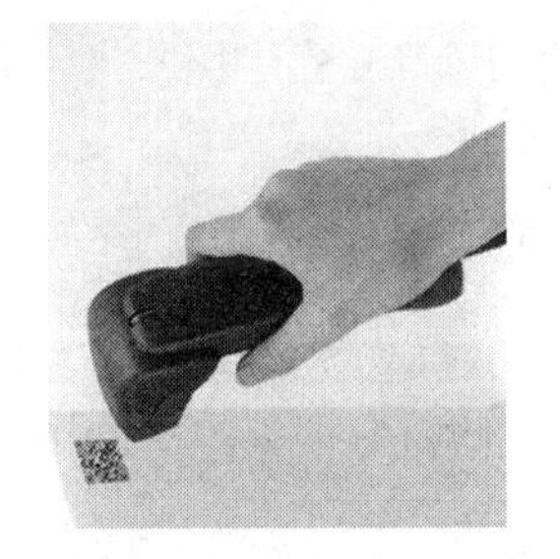

图 1-1-10　条形码阅读器

输出设备

输出设备以软拷贝或硬拷贝的形式输出结果。软拷贝是指呈现在显示器上的图形输出；硬拷贝是指通常由打印机输出的呈现在纸上的信息。因此，显示器和打印机是两种最常用的输出设备。

显示器是指具有类似电视机的显像屏幕的硬件。它使用阴极射线管（CRT）（图 1-1-11）、液晶显示器（LCD）（图 1-1-12）或其他图像显像技术将文本和图形呈现给计算机用户。阴极射线管价格便宜，显示清晰；而液晶显示器则相对较薄，占用空间较少。显示器的两个重要指标是尺寸和清晰度。尺寸是由显示器对角线的长度来表示，常见尺寸有 23 英寸、24 英寸、26 英寸和 27 英寸。清晰度通过分辨率表示，而分辨率又由像素决定。在显示器尺寸不变的情况下，像素越高，图像越清晰。显示器的常用品牌有三星、飞利浦、优派、LG、玛雅、明基、方正、美格等。

图 1-1-11　CRT

图 1-1-12　LCD

打印机是使用击打式打印技术或非击打式打印技术将显示器上的内容转移到纸上的设备。点阵打印机（图 1-1-13）因价格较低，一直是个人计算机上流行的打印机，它属于击打式的。最常见的非击打式打印机有喷墨打印机（图 1-1-14）和激光打印机（图 1-1-15）。相比之下，它们的噪音小，而且打印效果更佳。尽管与激光打印机相比，喷墨打印机可以打印彩色图像，然而，它的打印质量和速度却远远比不上激光打印机。打印机的著名品牌有佳能、爱普生、惠普、三星等。

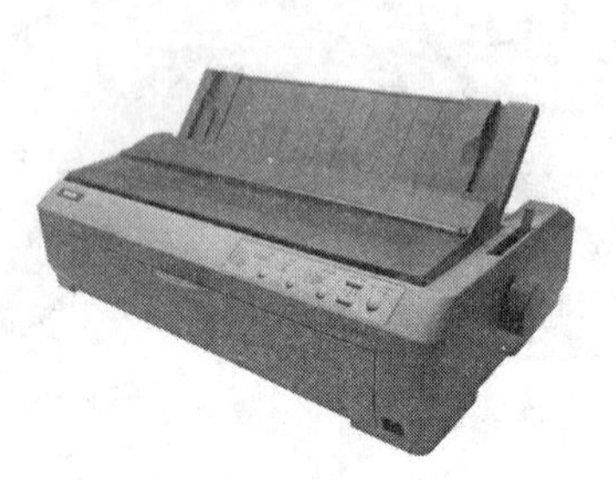

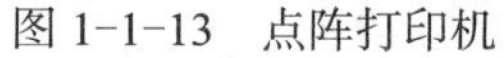

图 1-1-13　点阵打印机

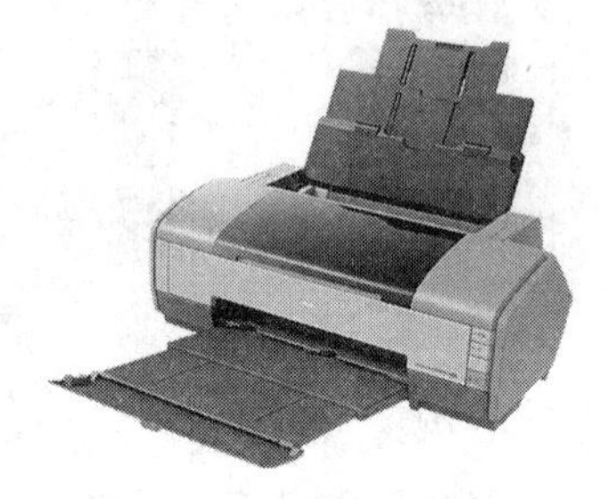

图 1-1-14　喷墨打印机

图 1-1-15　激光打印机

其他输出设备还有：用于输出音乐和讲话声音的音箱和耳机、为更大批观众呈现清晰信息的数据和多媒体投影仪。

输入/输出设备一体机

为节省空间和费用，或因为某种特殊用途，现在很多设备都将输入设备和输出设备融合在一起。常用的输入/输出设备一体机有传真机（图 1-1-16）、多功能设备（图 1-1-17）、网络电话、一体计算机（图 1-1-18）等。

图 1-1-16　传真机

图 1-1-17　多功能设备

图 1-1-18 一体计算机

1.2　主板和 CPU

本节介绍计算机硬件的重要组成部件，主要有主板、CPU 和内存。

主板

主板（图 1-2-1）是整个计算机系统的沟通网。它的作用就好比交通枢纽，计算机单元的每个部件都直接与其相连，部件之间的联系也因此得以实现。主板很重要，没有它，输入/输出设备，如显示器、键盘、打印机等就不能与计算机系统交互。

主板位于台式机箱的下面，它上面有几个插槽。各种计算机的卡，包括声卡和显卡，都能插进这些插槽里。卡是由芯片组成的电子元件，而芯片就是电路板。声卡控制计算机的声音，显卡在屏幕上显示图像。

图 1-2-1　主板

CPU

图 1-2-2　CPU

中央处理单元（图 1-2-2），即 CPU，是计算机的“大脑”。它位于微处理器的一个芯片上，微处理器通常位于插进主板的一个盒子里。CPU 读取并翻译软件指令，协调必须进行的处理活动。每个 CPU 都有一套独特的操作，如 ADD，STORE 和 LOAD 代表处理器的指令集。CPU 的设计影响到计算机的处理能力、速度和计算机能够有效利用的主存空间。如果你的计算机里的 CPU 设计得很好，就能够在很短的时间里完成十分复杂的任务。

CPU 有两个功能单元——控制单元和算术-逻辑单元。

- 控制单元

控制单元告诉计算机其他部分怎样去执行程序指令。在控制单元的控制下，从输入设备输入的程序和数据，被自动临时存储在内存里，然后得以执行，最后结果被输出并打印。程序由一系列指令组成，执行程序的过程就是按照一定顺序处理指令的过程。控制单元产生一系列控制信号，从内存读取指令，分析指令，执行其操作，然后再确定下一个指令的地址。这种逐步操作以惊人的速度反复重复，直到程序运行结束。

- 算术-逻辑单元

算术-逻辑单元，即 ALU，执行两种运算：算术运算和逻辑运算。算术运算根据算术规则进行基本的数学运算，如加、减、乘、除、确定绝对值等。逻辑运算就是进行比较。例如，将两个数据相比，看一个数据是等于（=）、小于（<）还是大于（>）另一个数据。如果两个数据相等，就继续进行处理；如果不相等，处理就结束。在计算机中，复杂的运算经常被分解成一系列算术运算和逻辑运算。运算中的两个数据被称为源运算数据，通常存在内存里，运算结果也存在内存里。

内存

内存是存储正在处理的程序和程序所使用的数据（包括运算结果）的地方。计算机运行指令时，先从内存取指令，然后执行。如果计算机要从内存再次取指令，就有必要再次访问内存。因此，内存的速度直接影响计算机的速度。

图 1-2-3　内存条

与 CPU 一样，内存也位于与主板相连的芯片上。有三类众所周知的内存条（图 1-2-3）：随机存取存储器（RAM）、只读存储器（ROM）和互补金属-氧化半导体（CMOS）。

• RAM

当谈到计算机内存时，通常指的是随机存取存储器（RAM），它是应用最广的内存。RAM存储计算机正在处理的程序和数据。在断电或关闭计算机时，RAM 中的所有内容就会丢失，所以说 RAM 是临时的或者短暂的存储器。正是由于此原因，当我们输入文档时，最好每隔几分钟就保存一次。

RAM 分两类：静态随机存取存储器（SRAM）和动态随机存取存储器（DRAM）。如果没有遇到断电，SRAM 中的内容会保留很长一段时间，而 DRAM 中的内容即使没有断电也会在一段时间（例如几毫秒）之后自动消失。因此，与 DRAM 相比，SRAM 使用起来更加方便，更加简单，其速度也更高。然而，SRAM 的容量小。

• ROM

只读存储器（ROM）存储程序。这些程序在制作芯片时就写入 ROM 芯片里了。也就是说，这些程序是固定的，用户只能使用却不能改变它们。计算机能够读取 ROM 芯片上的程序，却没法在 ROM 中写入任何信息，这就是为什么这类内存被称为只读存储器。

ROM 中的程序是一些特殊的指令。例如，按“POWER”键，计算机就能够自动启动；输入字母，字母就能够在屏幕上显示出来。这些都是 ROM 中指令的结果。可以看出，这些指令都是用于计算机运行的。

• CMOS

互补金属-氧化半导体 CMOS 芯片为计算机提供灵活性和扩展性。CMOS 中的信息是计算机启动时所要求的一些必要指令。与 RAM 不同，CMOS 是由电池供电。遇到断电或关闭计算机时，CMOS 中的内容不会丢失。与 ROM 中的固定程序也不一样，CMOS 中的程序可以加以改变。

1.3 存 储 设 备

内存和外存

什么是内存？什么是外存？它们之间有什么不同吗？是的，不同。了解“外存”和“内存”的不同很重要。当人们谈及计算机用到“内存”这个术语时，一般指被称为随机存储器或 RAM 的计算机主存储器，它由固定在主板上的芯片构成。内存有时被称为暂时存储器，因为如果断电，这部分记忆会丢失。相反，“外存”是指个人计算机所用的永久存储器，又称辅助存储器，通常是指个人计算机硬盘、CD 等。外存是永久的，因为断电后数据和程序还存在。

存储设备和存储介质

所有的存储系统都包括两部分：存储设备和存储介质。存储设备，如硬盘驱动器，CD 或 DVD 驱动器，它们都安装在机箱内，存储设备在存储介质上读写数据和程序。存储介质必须插入存储设备之后才能读写数据和程序。

存储介质——可移动介质和固定介质

在许多存储系统中，介质必须插入设备计算机才能读/写。这些称为可移动介质存储系统，例如：CD 和 DVD。另外，固定介质存储系统，如大多数硬盘驱动器系统，是把存储介质（硬盘）封在存储设备（硬盘驱动器）内，用户无法移动数据。

- 硬盘

几乎每台个人计算机都包含一个或多个硬盘驱动器系统，专门用来存储程序和大量的数据。这个系统在计算机的机箱内，不可移动。

硬盘驱动器系统是一个封闭的系统，在硬盘驱动器的盒子里，有一张或多张圆形的金属盘，金属盘上有许多磁道，这就是存储介质——硬盘。每个盘片需要两个读/写头，每面一个。这些读/写磁头被固定在一个称为读/写机构的设备上。当轴和盘片一起以每秒几千转的速度旋转时，读写机构在磁盘表面内外移动磁头来访问所需要的数据。

图 1-3-1　硬盘

和可移动介质相比，硬盘（图 1-3-1）存取速度更高、存储量更大、可靠性更强、成本更低。

然而，可移动介质系统有其他一些优势，包括下列几点：

无限存储容量——当一张盘满了，可以把一张新盘插入到存储设备中继续存储。

可移植性——可以很容易地与其他人或其他计算机共用一张盘。

备份——你可以把有价值的数据复制到可移动介质上，并把它单独保存起来。

安全性——对于敏感的程序或数据，可以把它存储在可移动介质上并保存在安全的地方。

- 软盘

软盘（图 1-3-2），有时又称磁盘，是可移动介质，而且很便宜，曾是个人计算机主要的存储介质之一。

有几种类型的软盘，分别有不同的容量。传统的软盘是 1.44 MB 的 3.5 英寸软盘，现在被高存储量的 U 盘所取代。

图 1-3-2　软盘

- U 盘

U 盘（图 1-3-3）是可移动、可重写的。U 盘由存储芯片组成，芯片包在一小片塑料内，U 盘的前端有一个 USB 接头。要访问存储在 U 盘上的数据，你只需把 U 盘插入计算机的 USB 端口。和软盘相比，U 盘更小、更轻、存取速度更快、存储密度更大，可以存储更多的数据，可靠性更强。U 盘的存储量从 8 GB 到 128 GB 或更大。U 盘已经变得日益普遍。

图 1-3-3　U 盘

• 光盘

图 1-3-4　光盘

光盘（图 1-3-4）通常称为高密度盘（CD），光盘用激光束读写数据。它的存储量比软盘要高得多。一张标准的 CD 可以存储高达 650 MB 的数据。一张 DVD 可以存储的信息量比一张 CD 还要多，因为 DVD 可以在两个面上存储信息。现在光盘被广泛应用，它是软件发行的标准形式，也被普遍用来存储高存储量的音乐和视频文件。

有几种类型的 CD 和 DVD：

CD-ROM（只读光盘）：不可以在磁盘上删除和添加数据。只能读取 CD-ROM 磁盘上的数据。

CD-R（可记录光盘）：可以在 CD-R 光盘上写数据，但只能写一次。

CD-RW（可重写光盘）：可以在 CD-RW 光盘上删除和重写数据。

DVD（数字通用光盘）：也包括 DVD-ROM 和可重写的 DVD，它是一种高存储量的光学存储格式，容量从 4.7 GB 到 17 GB。大多数的 DVD 驱动器可以播放计算机 CD 和音频 CD，但是 CD 驱动器不能播放 DVD。

第 2 章　计算机软件

一台计算机本是一个无生命的装置。它没有自身的智能，必须给它指令，它才能知道怎样去执行任务，这些指令就称为软件。计算机软件通常被分为两类：系统软件和应用软件。我们可以把系统软件理解为是计算机使用的软件，把应用软件理解为我们用户使用的软件。

2.1　系 统 软 件

系统软件介绍

系统软件是计算机管理其内部资源的软件。这种软件是运行基本的操作，告诉计算机硬件该做些什么、怎么去做、何时去做。系统软件不是单一的软件，而是几乎或者根本不需要用户介入的处理数百个技术细节的软件集合。系统软件分为四类：操作系统、实用程序、设备驱动程序和语言翻译器。

• 操作系统

人们总是希望充分利用计算机系统的所有资源，增强计算机系统的功能，为用户提供便利的工作环境，操作系统正是为此目的而发展起来的。操作系统是最重要的系统软件，是计算机不可或缺的部分。它与应用软件和计算机相互作用。操作系统的主要功能是工作管理、内存管理和设备管理。现在有很多种操作系统，应用最广泛的有 Windows，UNIX 和 Linux。其中几个将在后面讲到。

• 实用程序

实用程序执行与计算机资源管理有关的具体任务（图 2-1-1），目的是使计算机运算更加容易。计算机运行时，有可能出现任何问题。例如：病毒破坏了系统，致使计算机不能运转；硬盘受损，存储的信息有可能丢失等。而这些问题，都能通过实用程序加以解决。

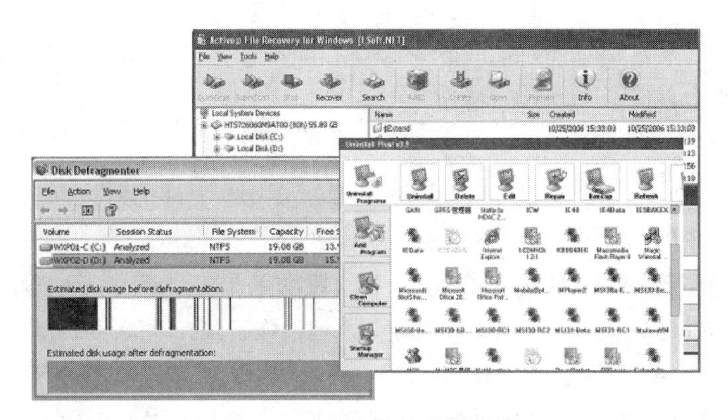

图 2-1-1　系统软件实用程序

不同的实用程序执行不同的任务。以下是经常使用的实用程序。

备份软件是为了防止原始文件在磁盘损坏时丢失或受损而对文件进行复制的软件。这种软件对数据进行压缩，以占据最小的空间。

文件压缩程序是压缩由于格式化而产生的松散空间。

杀毒软件是保护计算机系统不受侵入计算机内部的病毒的破坏。这些软件是必备软件，因为病毒随时都有可能传播。

磁盘碎片整理软件是找出并删除不需要的碎片，重新整理文件和未被使用的磁盘空间，使操作最优化。

卸载软件是从硬盘彻底而安全地删除不需要的程序或相关文件。

文件恢复软件是恢复已删除或被损坏的软件。

- 设备驱动程序

设备驱动程序是一种特殊程序，它和操作系统一起运行，使设备与计算机系统其余部分能够交流。每一种设备，如打印机、声卡、显卡、鼠标，都有其驱动程序，在计算机系统启动以后，操作系统就将驱动程序装入内存。设备驱动程序将操作系统或用户的命令转换成它所连接的设备能够理解的命令。

计算机系统装入一种新设备时，必须安装此设备的驱动程序，这样设备才能使用。如果我们手头没有某个驱动程序，可以从生产商网站上下载。

- 语言翻译器

低级语言和高级语言都称为程序设计语言。低级语言又称机器语言，由 0 和 1 组成，是计算机能够直接理解的最基本的语言类型。高级语言如 C，C++，Pascal，是由程序员编写的语言。高级语言要想被计算机理解并处理，必须先翻译成机器语言。语言翻译器就是将高级语言翻译成机器语言的软件。语言翻译器有三种：汇编程序、编译程序和解释程序。

Windows 操作系统

在过去几年里，微软公司开发了许多不同版本的 Windows 操作系统。最早的版本是 Windows 1.01，这个版本很简单。然后出现了 Windows 3.x，Windows 9.x，Windows NT，Windows 2000，Windows XP，Windows Vista，Windows 7，Windows 8，Windows 10 等版本。下面介绍其中几个版本：

Windows 3.x 发行于 20 世纪 80 年代和 90 年代初。Windows 3.x 代表这个软件的版号，例如

Windows 3.0、3.1 等。Windows 3.x 为 DOS 系统提供了图形用户界面。它使用了菜单、窗口和图标，代替了 DOS 的命令行。

1995 年，微软公司发布了一种新的用于个人计算机的 Windows 版本，称为 Windows 95，目的是代替 DOS 和 Windows 3.x 操作系统。Windows 95 的图形用户界面比以前的版本更加简单，而且图形用户界面不再是一个外壳，而被集成进了操作系统中。跟 Windows 3.x 一样，Windows 95 也使用窗口和桌面。它支持电子邮件、传真、多媒体、长文件名以及即插即用，这些都使得安装新硬件的过程更加容易。Windows 98 与 Windows 95 很相似，但增强了几个功能。Windows 98 集成了因特网支持技术，提供了附加的桌面用户界面定制命令，支持大硬盘，也支持 DVD 和 USB 接口技术。

Windows 2000 是 Windows NT 的升级版，它有几个不同的版本。事实上，Windows 2000 操作系统是一个系列产品，包含四个独立的组件，分别是 Windows 2000 Professional，Windows 2000 Server，Windows 2000 Advanced Server 和 Windows 2000 数据中心服务器版。

2009 年 10 月 22 日，微软公司发布了 Windows 7 版本。Windows 7 不同于其前身 Windows Vista，它引入大量新特征。它的用意更专注于视窗操作系统产品线的增值升级目标，其目标是用来兼容非 Vista 系统时期的应用和硬件。Windows7 具备多媒体处理功能，拥有重新设计的被称为超级任务条的视窗系统用户界面及家庭组网络系统功能，并且在性能上有所提升。

2012 年 2 月 29 日，微软公司发布了 Windows 8 用户体验版本。尽管启动屏幕还可以通过单击屏幕左下角的按钮或窗口触发，但是启动按钮是自 Windows 95 系统以来首次在任务栏上不再显示。微软总裁称自开发者版本发布以来已经修改 10 万多次。在发布的第一天 Windows 用户体验版已经超过 100 万次被下载。

2015 年 7 月 29 日 Windows 10 发布。Windows 10 不仅能用于计算机，也能用于手机和平板电脑。这一版本恢复了用户熟悉和喜欢的开始菜单。使用户能够很容易就找到文件和应用软件。Windows 10 能快速启动和重置，电池待电时间更长，能兼容所有设备和 Windows 应用软件。因此，Windows 10 既能用键盘，也能用触摸屏。

UNIX 和 Linux 操作系统

- UNIX 操作系统

UNIX 操作系统最早是由美国电话电报公司贝尔实验室的丹尼斯里奇和汤普森开发的。自 20 世纪 70 年代开发以来，UNIX 操作系统经由许多个人和公司，特别是加利福尼亚大学伯克利分校的计算机科学家的努力而得到增强。UNIX 有许多不同的版本。

众所周知，Windows 操作系统是专门为 Intel 芯片设计的，Mac 操作系统是专门为 Power PC 设计的。而 UNIX 操作系统很灵活，支持多种处理器架构。它在各类计算机系统，即从个人计算机到大型机上广泛使用。它还能通过网络连接很容易地整合不同生产商生产的设备。

UNIX 是一个多用户、多任务操作系统。它可以允许一到数百个用户同时运行不同的程序，还允许一个用户同时运行多个程序。

- Linux 操作系统

莱纳斯·托瓦尔兹是芬兰赫尔辛基大学的一名学生，他于 1991 年开发了 Linux 操作系统。Linux 是 UNIX 的兼容系统，支持很多软件。莱纳斯将此操作系统免费提供给大家，使其得到进一步发展。因为全部的 Linux 操作系统源代码是免费的，所以任何人都可以修改源代码来提高性能或者为某个特定应用进行修改。

Linux 操作系统具有 UNIX 和其他任何操作系统所具有的特点，如多用户、多任务、世界上最快的 TCP/IP 驱动程序等。它支持 32 位和 64 位多任务处理技术。Linux 操作系统还具有先进的网络技术，其网络支持优于其他大多数操作系统。

Linux 操作系统有着广阔前景。它不仅免费发布，而且它的功能性、适应性和稳健性都使其成为 UNIX 和微软操作系统的主要替代品。

2.2 应 用 软 件

应用软件介绍

当今，有上千种软件产品可供用户选择使用。这些软件产品执行大量不同的任务。计算机用户可以使用它们写信、发电子邮件、掌握财务状况、创建图表、学外语、做演示文稿以及许多其他应用。这些软件执行具体的任务，我们称其为应用软件。

应用软件分为两类：通用应用软件和专用应用软件。通用应用软件包括文字处理软件、电子表格处理软件和演示文稿制作软件等。专用应用软件包括许多其他软件，其功能很具体。多媒体和网页设计就是人们所熟悉的专用应用软件。

当我们使用应用软件时，必须了解以下功能。

窗口（图 2-2-1）是含有文档的矩形框。用户可以同时编辑一个或多个文档。每个文档位于一个独立的窗口中。当处理一个包括几个不同文件的大项目时，这一功能就会凸显出来。我们可以改变窗口的大小，移动和关闭窗口。

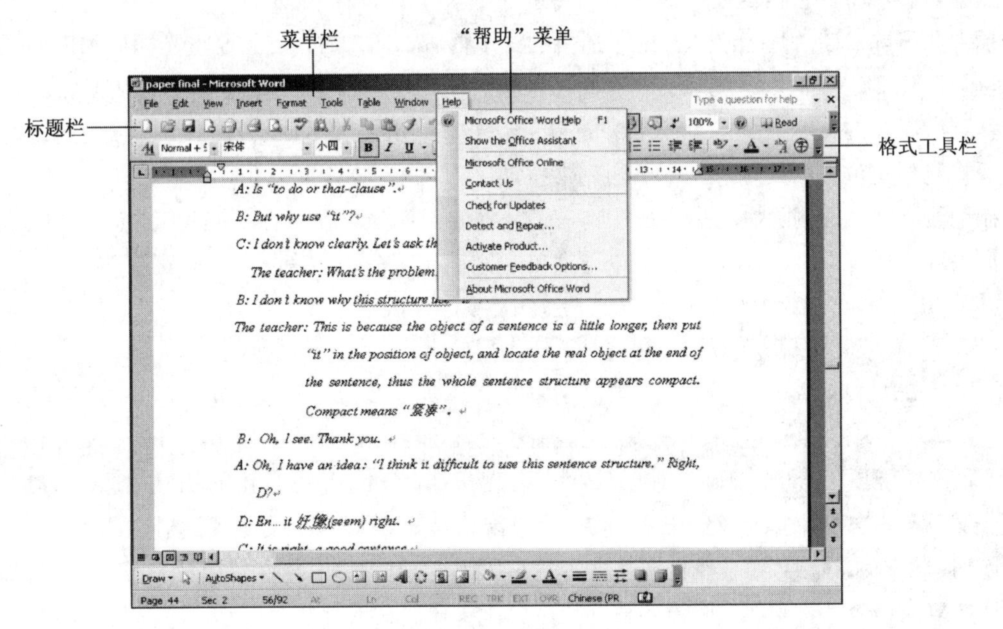

图 2-2-1　窗口

在计算机屏幕的顶部，有一个菜单栏。它会显示各个菜单，菜单下还包括一些命令。当选择一个菜单并单击，就会看到一个下拉菜单。下拉菜单中的选项即是与这个菜单相关的命令。

在菜单栏上，有一个命令是“帮助”。“帮助”菜单可提供一些帮助功能，解释怎样完成各种任务。

工具栏位于菜单栏下面，它们是一些经常使用的按钮。工具栏包括标准工具栏和格式工具栏。这些都有助于用户快速使用这些命令。

有时候在屏幕上还会出现一个对话框。对话框是用来收集用户的信息和显示帮助信息。

当输入信息时，需要空间。文本框就为我们提供空间从而输入一个数字、一个文件或文件夹的名字。在文本框中输入信息之后，按【Tab】键就会移动到下一个对话框。

文字处理软件

用文字处理软件可以生成各种类型的个人或商业信函，包括报告、通知、信件、备忘录、手稿以及其他形式的书面文件。人们可以建立、编辑、保存和打印这些文件。文字处理软件是最灵活、应用最广泛的应用软件工具之一。

如今，应用最广泛的文字处理软件有：用于个人计算机的 Microsoft Word 和 Corel WordPerfect；用于苹果机的 WordPerfect。

下面介绍文字处理软件的功能特征。

- 创建文档

使用文字处理软件生成新文档很容易。只需要使用键盘输入文本内容即可。所有文字处理软件都有换行功能。当输入内容时，一旦当前行输满了，换行功能会自动开始新的一行。但是，如果想另起一段或者留一空行时，就需要按【Enter】键。

- 编辑文档

当编辑文档时，根据需要可能会用到文字处理软件的下列功能。

插入和删除：插入是指给文档中添加新的文本内容。插入时，首先要将光标移到需要插入文本的位置，然后进行输入。删除文本有两种方法：一种是使用【Delete】键。使用此键时，先将光标移到所删除的文本前面，再按【Delete】；另一种是使用【Backspace】键。使用此键时，相反，要将光标定位在所删除的文本后面，再按键。

撤销删除：撤销删除命令指的是在删除了文本的一部分内容之后，又改变主意，可以恢复删除。如果连续做了好几次删除，就可以连续单击撤销删除命令，恢复所有的删除。

拼写和语法检查：这些功能有助于我们更好地输入。当键入文本内容时，文字处理软件会自动识别拼写错误，然后列出相似的字词以供选择。一些文字处理软件有自动改正功能，可以将一些普通的错误，如“adn”自动更正为“and”。但是，如果我们输入的字词不在文字处理软件的词典内，也会被标记为输入错误，如一些技术术语或人名。语法检查帮我们识别输入中的语法错误，如主谓一致、句子不完整、标点符号和大小写等问题。这些语法错误也会通过从建议的修改方法中选择正确的输入而加以改正。同类词典是文字处理软件的又一个功能。当查阅单词时，它会提供相关的词汇。

剪切、复制和粘贴：如果我们想将文本的一段或者一块从一个地方移到另一地方，就需要使用剪切或复制命令。首先，选中需要移动的部分，选择剪切或复制命令，然后将光标移到需要插入的地方，使用粘贴命令插入。

查找和替换：查找命令帮助我们在文档中查找字母、字或者词，而替换命令帮助我们用确定的文本内容去替换已查询到的文本。例如，我们能使用查找命令在文档中很快找到所有的

“White”，然后单击替换命令，用另一词“Black”替换掉“White”。

- 文档格式

文档格式指的是改变文档的外观。文档格式有许多选择。

首先，可以设置字符格式。用鼠标选中需要格式化的部分，然后使用合适的格式。我们可以按照需要或者自己的愿望确定字体、字号和字形。字体指的是使用具体设计的字符集，如 Times New Roman，Arial Black。字号是字符的大小，单位是磅。例如，我们可以将一篇文章的题目设置为 14 磅的 Times New Roman。我们还可以决定是否要将字符加下画线、加粗或者倾斜。

其次，可以设置段落格式（图 2-2-2）。常用的段落格式包括行距、间距、对齐和缩进。行距是指行与行之间是单倍行距还是双倍行距，或者其他。间距是段前和段后的空间距离。段落对齐有五种类型：两端对齐、居中、左对齐、右对齐和分散对齐。两端对齐是指文本的左右两边页边距相等。居中是文本的每一行都处在页面的中间。左对齐是文本的左侧在一条线上，而右对齐是文本的右侧在一条线上。分散对齐是指段落中各行的字符等距离排列在左右边界之间。缩进包括段落的首行缩进和悬挂缩进。

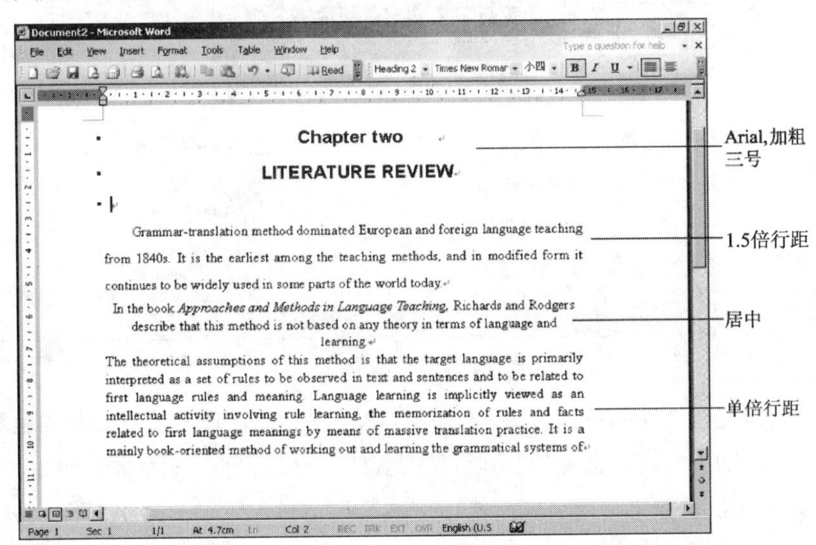

图 2-2-2　格式化段落

大多数文字处理软件都有页面设置功能。用户可以在页面顶部或底部设置页码，也可以添加页眉和页脚。页眉是自动打印在每页顶部，通常是日期、文档题目，或者诸如此类的文本。页脚是打印在每页底部。我们还可以格式化文档。事实上，当我们对文档的所有页面进行格式化时，我们就称之为格式化文档。另外，可以给文档添加背景，生成尾注和目录表，或者把其他文件的图片加入文档中来。

一些文字处理软件还有下述功能：允许用户在因特网上访问 Web 网页，搜索信息，设计并发布网页，在文字处理文档中插入一个指向某网页的超链接，还可以从其他网页中检索图片。

电子表格处理软件

电子表格处理软件被人们广泛用于任何行业。人们用它来组织和分析数据、建立图表。如今，使用最广泛的电子表格处理软件是微软 Excel, Lotus 1-2-3 和 Corel Quattro Pro。

下面介绍电子表格处理软件的功能特征。

电子表格是由行和列形成的矩形方框。行由数字来标记，而列由字母标记。单元格是行与列交叉生成的，数字、词或词组、公式就存储在单元格里。数字和公式称为数字型输入，词和词组称为文本型输入。

公式是指运算指令。电子表格处理软件提供了丰富的环境来建立复杂的公式。利用一些数学运算符和单元格填写规则，就能进行大量的计算。例如，公式 B4+B5+B6 表示将单元格 B4、B5 和 B6 中的值相加。如果将 1、2 和 3 分别输入单元格 B4、B5 和 B6 中，然后选中空单元格 B7，输入公式= B4+B5+B6，按【Enter】键，单元格 B7 则显示值 6（图 2-2-3）。

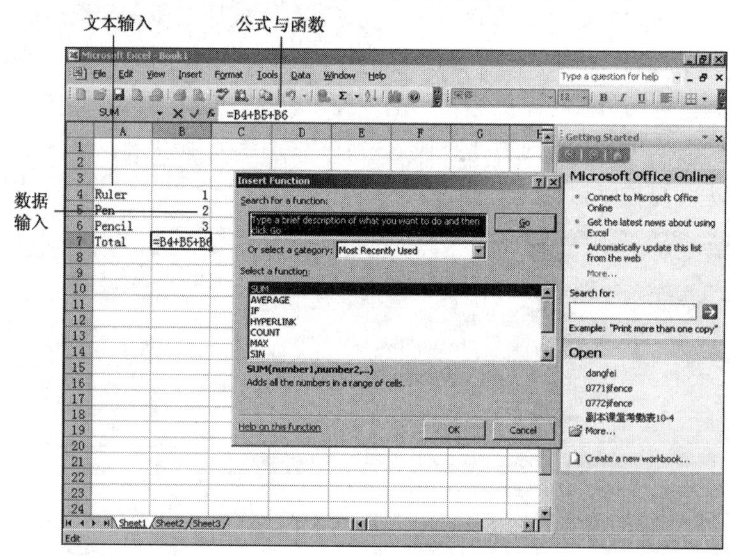

图 2-2-3　求和操作

函数是电子表格处理软件的特殊工具。函数能自动、快速而简便地进行复杂的运算。事实上，函数是已经写入的自动进行某些运算的公式，是频繁使用的公式的速记版本。例如，假设要将单元格 A1 到 A5 的值相加，如果使用公式，就必须输入=“A1+A2+A3+A4+A5”；如果使用函数，就输入“=SUM (A1:A5)”。显然，SUM()函数比公式要短得多，这就使工作更加容易。

电子表格处理软件可以进行重新计算，即如果我们改变了电子表格中一个或多个数字，所有相关的公式将重新计算，产生新的结果。例如：给单元格 A1 中输入 10，A2 中输入=A1+5，按【Enter】键，A2 中显示 15。如果用 5 替代 A1 中的 10，选中 A2，A2 中的值自动变为 10。这一简单的功能会大大减少我们数小时的繁重工作。

电子表格处理软件还有一个功能，即在其标准工具栏上有自动求和按钮。自动求和按钮是把求和函数插入某一单元格里并且建议一个求和区域。如果建议的求和区域不正确，可以拖动鼠标以选定正确区域，再按【Enter】键。求和按钮还有一个菜单，打开菜单可发现平均值、计数、最大值和最小值函数。此外，它还包括“其他函数”命令，选择此命令会打开“插入函数”对话框，在这里可以选取任何函数功能。

电子表格处理软件的另一功能是图表向导。图表向导帮助我们生成图表。图表能直接显示工作表中的数据，这很容易理解。图表类型很多，其中最常用的有饼图、折线图和柱形图。股价图可帮助人们分析股票市场的情况（图 2-2-4）。

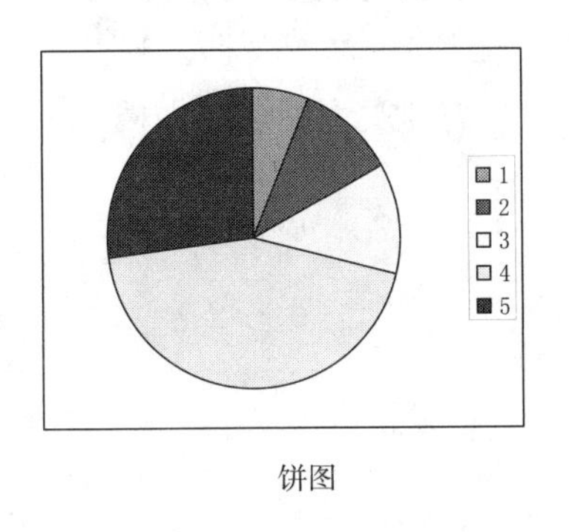

饼图

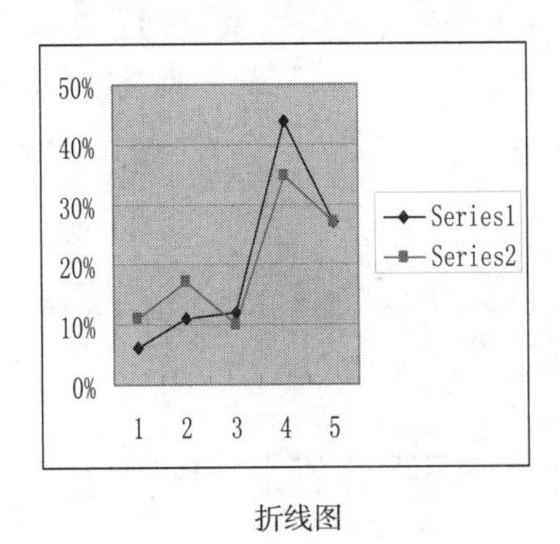

折线图

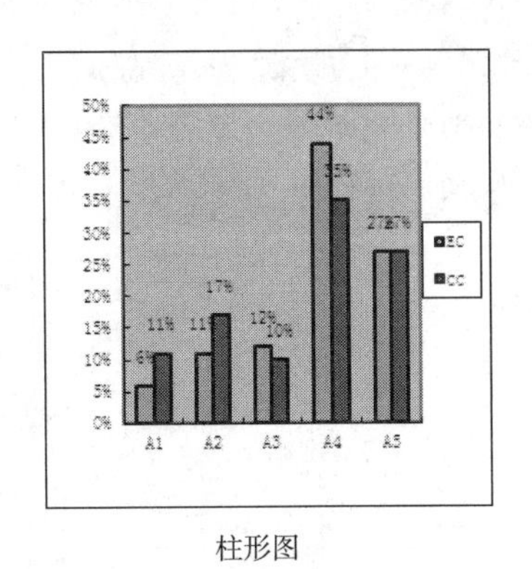

柱形图

图 2-2-4　图表类型

使用电子表格处理软件，还可以在工作表上做一个超链接，用来在内部网上或因特网上访问其他文件。我们还可以生成并使用查询功能，从网页上检索信息并直接插到工作表中。

演示文稿制作软件

演示文稿制作软件就是用可视的物体——图片和信息来创建演示文稿制作。演示文稿制作看起来很有趣，很吸引人，被广泛应用于商业及课堂中。许多人使用演示文稿制作软件来做报告、演讲、发布信息等等。如今，最常用的演示文稿制作软件有 Microsoft PowerPoint, Lotus Freelance Graphics 和 Aldus Persuasion。

下面主要介绍 Microsoft PowerPoint 的功能特点。

当打开 PowerPoint 窗口时，我们会看到一张幻灯片。这是第一张幻灯片，称为“标题幻灯片”。标题幻灯片中有加标题和副标题的位置框。如果不需要副标题，可以不去管它，演示文稿中也不会出现。当单击“单击添加标题”位置框后，就可以输入文本（图 2-2-5）。要添加新幻灯片，就需要单击“新幻灯片”按钮。每次单击“新幻灯片”按钮，都会出现一个默认幻灯片。

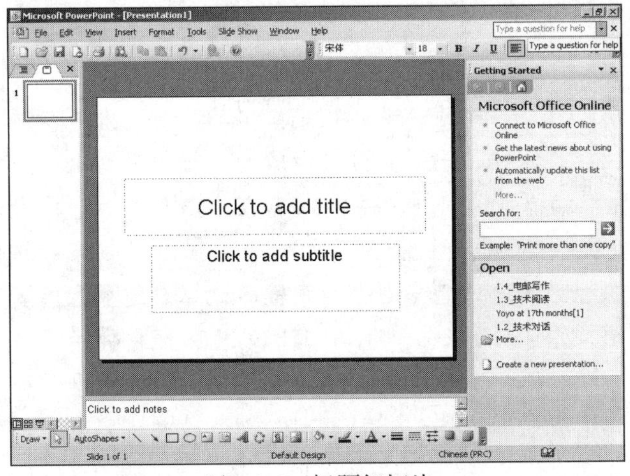

图 2-2-5　标题幻灯片

当浏览幻灯片时，单击菜单栏中的“视图”，可选择“幻灯片浏览”或者“幻灯片放映”。另外一种方法就是普通浏览。打开 PowerPoint，我们看到的就是这种默认浏览方式。还有一种方法是大纲浏览，这种方法能使整个演示文档以大纲的形式进行文本表述。

Microsoft PowerPoint 具有轻松创建动态的演示文档功能，我们也称其为动画。它包括特殊的视觉和听觉效果，比如有图片、电影、声音。有各种各样的模板帮助设计演示文档。模板就是预先设置好的文件。这些预先定义好的设置可用作模式来创建许多普通类型的演示文档。

第3章 程序设计语言

3.1 程序设计语言简介

程序是指导计算机完成所要求的数据处理任务的一系列指令和语句。有许多种程序设计语言可用来编写计算机程序。传统上将计算机程序设计语言划分为四个级别，分别是：机器语言、汇编语言、高级语言、面向对象的程序设计语言。

机器语言

机器语言是最低级的程序设计语言。它是计算机唯一能识别并直接执行的语言。换句话说，实际上计算机唯一能执行的程序指令是用机器语言写的指令。每一条机器语言指令对应一串二进制代码。所以，我们可以说，机器语言是一种由二进制代码指令所组成的、由计算机直接使用的语言。

每一条机器语言指令只能完成一个很低级的任务。处理过程中的每一小步在机器语言中都必须明确地用代码写出。即使是把两个数加在一起这样的小任务也要用三条二进制代码指令，程序员必须记住哪个二进制数的组合对应的是哪一条指令。所以，机器语言很难掌握和使用。而且，因为机器语言指令是一串二进制代码，可读性差，不易记忆，容易出错，程序的调试和修改难度也很大。事实上，现在几乎没有程序是用机器语言写的。取而代之，大多数程序是用更高级的语言写的，然后把它翻译成机器语言，再由计算机执行。

汇编语言

汇编语言出现于 20 世纪 50 年代初期。它是最早的开发出来可以帮助编程人员的工具。汇编语言是由一些比较容易识别和记忆的符号组成的。每条机器语言指令在汇编语言中都有等价的命令。例如，在汇编语言中，语句“MOV A, B”命令表示计算机把数据从一个单元复制到另一个单元。而机器语言中同样的指令是由一串 16 位的 0 和 1 组成的。由汇编语言编写的程序称为汇编语言源程序，这种汇编语言的源程序计算机是不能直接执行的。一旦汇编语言的源程序写好，必须首先通过另一个叫作汇编器的程序把它翻译成机器语言，然后才能执行。与机器语言相比，汇编语言速度更快、功能更强。但它仍然难以利用，因为汇编语言指令是由一系列抽象代码组成的。另外，汇编语言的一个最大缺陷在于它是面向机器的。不同的 CPU 使用不同的机器语言，因此有多少种 CPU，就有多少种汇编语言、多少种编译器。

高级语言

高级语言出现于 20 世纪 50 年代中期，它和人类的思考及交流方式更接近了。高级语言是一种用各种意义的"词"和"数学公式"按照一定的"语法规则"编写程序的语言。如 C 语言、Basic 语言和 Fortran 语言都是高级语言。

一旦程序员用高级语言或汇编语言编写程序，这些程序必须转换成机器码。正如前面所提到的，用汇编语言编写的程序输入汇编器，把汇编语言的指令翻译成机器码，然后机器码作为汇编器的输出才能执行。对于高级语言来说，我们用其他的软件工具来帮助完成这个翻译过程。把高级语言的程序翻译成机器码的程序称作编译器。高级语言程序被编译，而汇编语言程序被汇编。

高级语言编写的程序可以在任何一台计算机上运行，但这台计算机要有合适的语言编译器。注意，编译器是程序，因此，编译器的一个机器码版本一定是适用于某一特定的机器来编译程序。因此，每一种高级语言必须有许多个编译器，来用于多个不同种类的机器。

高级语言有许多优点。例如，比汇编语言更易于学习；书写程序所需时间较少；提供较好的文档；易于维护；能够熟练使用这种语言编写程序的程序员将不受某一机器类型的限制；等等。

面向对象的程序设计语言

面向对象的程序设计语言是以传统的高级语言为基础，但是它们能使程序员按照组合对象集方式而不是指令列表方式来进行思考。对象是能够被看见、触摸到或感觉到的某种事物，对象的类型包括人、地点、物或事物。对象有属性，属性是描述对象令人感兴趣的特征的数据。以顾客这一对象为例，其属性可能有 CUSTOMER NUMBER（顾客编号）、CUSTOMER NAME（顾客姓名）、HOME ADDRESS（家庭住址）、HOME PHONE（家庭电话）等。使用面向对象的程序设计（OOP）、技术、程序员可以在对象之间建立关系，如一个对象可以继承其他对象的属性。与面向过程的程序设计技术相比，OOP 的主要优点是、程序员可以创建程序模块，在新对象加入时，不必修改。程序员只要创建一个新的对象，让它继承现有对象的某些属性即可。所以，面向对象的程序很容易修改。这也简化了程序员的工作，从而导致更多既可靠又高效的程序产生。要实现面向对象编程，我们需要使用面向对象的程序设计语言（OOPL），例如 Microsoft Visual Basic，C++, C，Smalltalk，Eiffel，Common LISP Object System（CLOS），Pascal，Java 和 Ada95 等。

3.2 C

C 是一种极简化的编程语言，是绝大多数计算机人员用来开发软件产品的语言。C 最初发展始于 1969 年，之所以叫作 C 是因为它的许多属性都继承自之前一种叫作"B"的语言。BASIC 曾经是最主要的微型计算机编程语言，20 世纪 70 年代末期 C 取代 BASIC，成为新的最主要的微型计算机编程语言。C 语言是一种编译型语言，也就是说，程序编好后，必须通过 C 编译器把程序转换为计算机可运行的程序。C 程序是一种人可以阅读的程序，但是由编译器编译的程序是计算机可读的和可执行的形式。有时候人们称 C 为"高级语言"，而实际上它比各种汇编语言仅高级一些。但相比汇编语言，C 语言有以下优势：第一，促进了模块化程序设计，每个程序块完成仅一个功能，把这些模块连接到一起就可构成程序。第二，代码读/写通常要容易得多，尤其是编写较长的程序时，这种优势更加明显。第三，C 语言更加灵活。由于当前大多数计算机系统都配置了 C 编译器及 C 库，它可移植到任何体系结构的计算机中。也就是说，C 最突出的特点就是它不与任何特定的硬件或系统绑定，这使得用户编制的程序不经过任何改动就能很容易地在几乎所有计算机上运行。事实上，编译器、库及其他更高级语言的解释器都是用C实现的。因此，C的可移植性至少不比其他更高级的语言差。

比较新的、面向对象的C语言被称为C++语言。C++是C语言的加强版，是20世纪80年代开发出来的。它与C语言兼容，也可以说是它的超集。所以，现有的C语言代码可以组合到C++程序中。C++包括C的基本特性，所有C++程序能被C编译器识别。C++是基于面向对象程序设计的概念，给程序提供基于现实生活的数据模型。因此，C++具有另外一些特性，如对象、类和面向对象程序的其他元素。对象的定义要经过它的类。类决定一个对象的所有方面，而对象是类的一个单个的实体。例如，可以从"人"类中建立一个对象"中国人"。"人"类定义什么是"人"对象以及"人"对象行为的所有相关信息，这个对象具有头发颜色、眼睛颜色、肤色等属性，还具有使它运作的方式，如说话、吃饭等。C++ 完全支持面向对象的程序设计，包括面向对象开发的四个特征：封装、数据隐藏、继承性和多态性。此外，C++还拥有许多与面向对象程序设计无关的新特性和新改进。C++也有可视化版本 VC++。所有这些表明，C++是用于图形应用的最通用的程序设计语言之一。

C的最新版本是C#，是为提高开发Web应用软件效率而设计的一种面向对象的程序设计语言。

3.3 Java

Java是由Sun Microsystems公司于1995年5月推出的Java程序设计语言和Java平台的总称。但是在本节中提到的Java，主要是指Java程序设计语言。

Java语言是一个支持网络计算的面向对象的程序设计语言。它最初来源于Sun公司1991年的"绿色计划"。以James Gosling为首的绿色团队共有13名成员，他们原先的目的是为家用消费电子产品开发一个分布式代码系统，这样可以把email发给电视机、冰箱这样的家用电器，从而对它们进行控制。开始，他们准备采用C++，但C++太复杂，且安全性差，最后他们基于C++开发出了一种新的语言 Oak，这是因为它的创造者 James Gosling 所在的公司窗外有一棵橡树（Oak）。后来他们发现已经有一种程序设计语言叫作Oak，所以当该团队的成员在一起喝一种叫Java（爪哇）的咖啡时，一个组员建议把它叫Java，这得到了大家的一致同意。但是，"绿色计划"却遇到了麻烦，因为消费性电子产品市场并未像人们预期的那样发展起来。幸好，1993 网络迅速发展起来，Sun公司发现Java可以用来创造动态网络主页，这给该计划带来了生机。1995年，Sun公司正式宣布了Java。

Java的主要发布版本和发布日期如下：

- JDK 1（1996年1月21日）
- JDK 1.1（1997年2月19日）
- J2SE 1.2（1998年12月8日）
- J2SE 1.3（2000年5月8日）
- J2SE 1.4（2002年2月6日）
- J2SE 5.0（2004年9月30日）

- Java SE 6（2006 年 12 月 11 日）
- Java SE 7（2011 年 7 月 28 日）
- Java EE 7（2013 年 10 月 27 日）

如今，Java 技术被广泛用于网络和各种产品，从因特网和巨型计算机到笔记本式计算机和手机，从华尔街市场模拟装置到家庭游戏机和信用卡，几乎无所不能。确切地说，它支撑着 45 亿多产品。

为什么 Java 有如此大的魅力呢？这是因为 Java 具有如下特征：简单性、面向对象、分布式、解释性、稳定性、安全性、可移植性、平台无关性、多线程。

简单性：Java 最初是为家用电器进行集成控制而设计的一种语言，因此它必须简单明了。Java 语言的简单性主要体现在以下三个方面：首先，Java 的风格类似于 C++，很容易被 C++程序员掌握；其次，Java 摒弃了 C++中容易引发程序错误的地方，如指针和内存管理；再次，Java 提供了丰富的类库。

面向对象：Java 语言的设计完全是面向对象的，它不支持类似 C 语言那样的面向过程的程序设计技术。

分布式：Java 支持 B/S 计算模式，因此支持数据分布和操作分布。数据分布是指数据可以分散在网络的不同主机上，而操作分布是指把一个计算分散在不同主机上处理。

平台无关性：Java 运行于 Java 虚拟机，实现不同平台的 Java 接口。用 Java 写的应用程序不用修改就可在不同的软/硬件平台上运行，即它是一种写一次，就可在任何机器上执行的语言。

多线程：Java 的多线程功能使得在一个程序里可以同时执行多个小任务。线程有时也称小进程。它是一个大进程里分出来的小的独立的进程。多线程的一大好处是更好的交互性能和实时控制性能。

与 C++相比，Java 的安全性更高，而且它去掉了 C++语言的许多功能，如指针运算、结构、typedefs 等，需要释放内存，这让 Java 的语言功能很精练。

如果需要了解 Java 的更多信息，可以登录 Sun 公司的 Java 网站 http://java.sun.com/ 或 http://www.javasoft.com/。如果需要中文的 Java 信息，则可以到清华和中科院的 FTP 上查找。

补充知识：

目前 Java 有三个版本，每个版本都能满足一些特定的开发需求。Java SE （原来的 J2SE）——Java 平台标准版定位于客户端，主要用于桌面应用软件的编程；Java EE （原来的 J2EE）——Java 平台企业版，定位在服务器端 Java2 的企业版，主要用于分布式的网络程序的开发，如电子商务网站和 ERP 系统；Java ME（原来的 J2ME）——Java 平台 Micro 版，主要应用于嵌入式系统开发，如手机和 PDA 的编程。

3.4 Visual Basic

Basic 语言是 20 世纪 60 年代美国 Dartmouth 学院的两位教授设计的计算机程序设计语言，

其含义是初学者通用的符号指令代码。Basic 提供使人愉快的友好的语言设计环境，因此简单易学，使用方便，很快就成为最流行的程序设计语言之一。20 世纪 80 年代，为了满足结构化程序设计的需要，新版本的 Basic 语言增加了新的数据类型和程序控制结构，其中较为人所熟悉的有 True Basic，Quick Basic 和 Turbo Basic。

1988 年，Microsoft 公司推出了 Windows 操作系统。图形用户界面的出现在计算机世界引发了一场革命，它深受计算机用户的欢迎。然而，对程序员来说，开发一个基于 Windows 环境的应用程序工作量非常大。可视化程序设计语言正是在这种背景下应运而生。可视化程序设计语言是编写程序的一种方法，使用代表一般程序设计规则的图标。

Visual Basic 是 Microsoft 公司开发的面向对象的程序设计语言，它是一种工具，Microsoft 公司的 Windows 操作系统用户用他来开发自己的图形用户界面应用软件。1991 年，Microsoft 公司首次发行 Visual Basic，然后又有其他版本如 Visual Basic 2.0 版……6.0 版，其最新版本为 VB.NET。20 世纪 90 年代中期，Visual Basic 十分流行。

Visual Basic 具有以下主要的功能特点。

Visual Basic 使用面向对象的程序设计方法，它把程序和数据作为对象，每一个对象都是可视的。程序员根据已经定义好的部件建立一个图形用户界面，通过描述部件对各种事件的反应来自定义部件。例如一个按钮，程序员要描述当这个按钮被单击之后会出现什么情况。因此，程序员只需编写对象要完成的事件过程的代码，从而提高程序设计的效率。

Visual Basic 是一种事件驱动的程序设计语言。事件驱动的编程方式非常适合图形用户界面。Visual Basic 使得用户通过绘图和安排用户的元件来快速地设计用户界面，因此，节约了大量重复任务的时间。

Visual Basic 提供了易学易用的应用程序开发环境，它具有丰富的数据类型，使用结构化的程序设计语言。

Visual Basic 的另一功能特点是 Active 技术。Active 技术允许用户使用由其他应用程序所提供的功能，例如由 Microsoft Word，Excel 和其他 Windows 应用程序提供的功能。Visual Basic 使程序员能够开发集声音、图像、动画、文字处理、电子表格和 Web 等对象于一体的应用程序。

Visual Basic 因特网的功能，使用户可以通过因特网或者因特内联网，很容易从用户的应用程序访问文档和其他应用程序，或者创建因特网服务器应用程序。

Visual Basic 是一种程序设计语言，它被认为是用来建立应用软件的快速应用开发软件。其他流行的程序设计语言还有 C++、HTML、SQL 等。

第4章 数 据 库

4.1 数据库简介

当今，人们生活在一个信息社会，利用信息系统来管理，从体育的统计数字到薪金的一切数据。同样，收款机和自动取款机也是由大型的信息系统所支持的。它们几乎影响到我们生活的方方面面。其中一个最流行的普遍应用的信息系统就是数据库管理系统。

数据库可以简单界定为一个结构化的数据集。数据库管理系统（DBMS），有时又称数据库管理者，是一组程序，它允许一个或多个计算机用户创建数据库和访问数据库中的数据。

设计数据库系统的目的是为了管理大量信息。为了达到这样的目的，人们在数据库中设计了复杂的数据结构用来表示数据。由于许多数据库系统的用户对于计算机的数据结构没有太深的了解，数据库开发人员经常通过如下几个层次向用户屏蔽其复杂性，以简化用户与系统之间的交互：

物理层。最低层次的抽象，描述了数据实际上是怎样存储的。物理层详细描述复杂的底层数据结构。语言编译器为程序设计人员屏蔽了这一层的细节。

逻辑层。比物理层层次稍高的抽象，描述了数据库中存储什么数据以及这些数据间存在什么关系，因而整个数据库通过少量相对简单的结构来描述。虽然简单的逻辑层结构的实现涉及复杂的物理层结构，但逻辑层的用户不必知道这种复杂性，逻辑层抽象是由数据库管理员所使用的，管理员必须确定数据库中应该保存哪些信息。程序设计人员也是在这个层次上工作的。

视图层。最高层次的抽象，但只描述整个数据库的某个部分。尽管在逻辑层使用了比较简单的结构，但由于数据库的规模巨大，所以仍存在一定程度的复杂性。数据库系统的多数用户并不需要关心所有的信息，而只需要访问数据库的一部分。视图抽象层的定义正是为了使用户与系统的交互更简单。系统可以提供同一数据库的多个视图。除了屏蔽数据库的逻辑层细节以外，视图还提供了防止用户访问数据库某些部分的安全性机制。例如，银行的出纳员只能看见数据库中关于客户账户信息的部分，而不能访问涉及员工工资的信息。

设计数据库系统的目的是为了管理大量信息。数据管理既涉及信息存储结构的定义，又涉及信息操作机制的提供。除此之外，数据库系统还必须保证当系统崩溃或企图进行非法存取时信息的安全性。如果数据被几个用户共享，则系统必须能够避免出现异常的结果。

4.2 关系型数据库

数据库是一组结构化的记录或数据的集合。计算机数据库系统依靠软件组织和存储数据。这些软件把数据库结构模式化成所谓的数据库模型。现在应用最普遍的模型是关系型数据库模型。还有其他一些模型，如层次模型，网络模型和面向对象的数据库模型。

关系、网络、层次及面向对象这些词指的是 DBMS 内部组织信息的方式。内部组织方式的不同会影响提取信息的速度和灵活性。

关系模型用表的集合来表示数据和数据间的联系。每个表有多列，每列有唯一列名。关系数据模型用一种称为关系的简单的二维表来表示数据库中的所有数据。只要两张表有相同的数据元素，即可将存储于一张表中的数据与另一张表中的数据联系起来。

例如，假定有一个如表 4-2-1 所示的数据库表，它包含一些电影信息。表中的每一行对应一条记录。表中的每一列对应一个字段。每一条记录都有相同的字段，其中存储着不同的字段值。也就是说，每一条记录都有包含特定数据的“电影编码”“标题”“类型”和“分级”。数据库表要有表名，如本例中的“电影”表。

表 4-2-1　电影表

电影			
电影编码	标　题	类　型	分　级
101	第六感觉	恐怖惊悚片	PG-13
102	回到未来	喜剧冒险	PG
103	怪物公司	动画喜剧	G
104	梦幻之地	幻想戏剧	PG
105	外星人	科幻恐怖	R
106	不死劫	惊悚片	PG-13
107	X 战警	动作科幻	PG-13
5022	伊莉莎白	戏剧	R
5793	独立日	动作科幻	PG-13
7442	野战排	动作戏剧战争	R

通常，表中一个或一个以上的字段被定义为关键字段。关键字段可以唯一确定表中所有记录中的一条记录。也就是说，表中每条记录的关键字段中所存储的值必须是唯一的。在“电影”表中，逻辑上会选择“电影编码”字段为关键字段。这样的话，两个电影可以有同样的标题。在这个例子中，当然“类型”和“分级”字段作为关键字段不合适。

在表 4-2-1 的“电影”表中，“电影编码”的字段值恰好是按递增的顺序显示的，但是它还可以按其他的方式显示，比如按电影标题字段的字母排序。本例中，表中每一行的数据之间没有内在的联系。关系型数据库表表现的是数据的逻辑视图，与底层的物理结构（记录如何在磁盘上存储）没有关系。

假设我们要创建一个电影租赁公司，除了需要列出要租赁的电影，还必须创建一个包含客户信息的数据表。表 4-2-2 这个客户表就呈现了这样的信息。和我们的电影表格一样，客户表格也包含一个客户号字段作为关键字段。

电影表格和客户表格分别在各自的表格中显示了各条记录是如何被组织起来的。然而关系型数据管理系统的真正强大在于它可以从概念上创建一个把各个不同的表格联系在一起的表格。数据管理系统通过各个表格中的共有字段把各个不同的表格连接在一起。关系型数据库这个名字就暗示了这个软件可以通过各个表格中的共同字段把各个不同表格中的数据联系起来。

表 4-2-2　客户表

客户			
客户编码	姓　名	地　址	信用卡号
101	Dennis Cook	123 Main Street	1234 5678 9876 5432
102	Doug Nickle	456 Second Ave	5678 9876 5432 1234
103	Randy Wolf	789 Elm Street	9876 5432 1234 5678
104	Amy Stevens	321 Yellow Brick Road	5432 1234 5678 9876
105	Robert Person	654 Lois Lane	1122 3344 5566 7788
106	David Coggin	987 Broadway	9988 7766 5544 3322
107	Susan Klaton	345 Easy Street	2345 6789 8765 4321

继续看电影租赁的例子，我们需要表示某一客户租赁某一电影的情况。因为“租赁”表是客户和电影之间的关系表，因此把它们表示在一条记录中。出租日和到期日是关系表的属性（字段有时又称做属性），也应该写在同一条记录中。表 4-2-3 中的“租赁”表包含了一系列表示当前出租的电影记录。

表 4-2-3　租赁表

租　赁			
客户编码	电影编码	出租日	到期日
103	104	3-12-2006	3-13-2006
103	5022	3-12-2006	3-13-2006
105	107	3-12-2006	3-15-2006

“租赁”表包含了关系（客户和电影）的对象（记录又称对象或实体）的信息，还有关系的一些属性。然而，它并不包含有关客户和电影的所有数据。在关系型数据库中，我们尽可能避免数据的重复。例如，在“租赁”表中不需要存储客户的姓名和地址——这些数据已经存储在“客户”表中。“租赁”表和“客户”表拥有共同字段“客户编码”。所以它们通过共同字段建立了关联。当需要相关数据时，我们用存储在“租赁”表中的“客户编码”来查询“客户”表中客户的详细数据。同样，“租赁”表和“电影”表通过共同字段“电影编码”建立了关联。当需要被租赁的电影数据时，我们用存储在“租赁”表中的“电影编码”在“电影”表中查询。

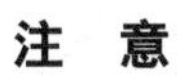

注　意

在“租赁”表中，“客户编码”的字段值 103 在两条记录中出现。这表明同一位客户租赁了两部不同的电影。

按照需要可以在不同的数据库表中对数据进行修改、增加和删除。当对库存的电影进行增加或删除时，就更新了“电影”表的记录。当有人成为新客户时，就在“客户”表中把他们加进去。根据不断的发展变化，我们随着客户租走和还回电影光盘，在“租赁”表中增加和删除记录。

4.3　数据库语言

数据库系统提供两种不同的语言：一种用于定义数据库的模式；另一种用于数据库的查询和更新。

数据定义语言（DDL）

数据库模式是由一系列定义来确定的，这些定义由称作数据定义语言（DDL）的一种特殊语言来表达。DDL 语句的编译结果是一系列表，这些表存储在一个称作数据字典或数据目录的特殊文件中。

数据操纵语言（DML）

数据操纵的意思是指：

对存储在数据库中的信息进行检索。

向数据库中插入新的信息。

从数据库中删除信息。

修改数据库中存储的信息。

数据操纵语言（DML）是一种用户可以访问和处理由适当的数据模式组织起来的数据的语

言。通常有两类数据操纵语言：

过程化 DML 要求用户指定需要什么数据以及如何获得这些数据。

非过程化 DML 只要求用户指定需要什么数据，而不必指明如何获得这些数据。

通常非过程化 DML 比过程化 DML 易学易用。然而，由于非过程化 DML 的用户不必指明如何获得数据，导致这种语言产生代码的效率不如过程化语言的高。我们可以通过各种优化技术来解决这个问题。

结构化查询语言（SQL）

结构化查询语言（SQL）是一种管理关系型数据库的综合性的数据库语言。它包括定义数据库结构的语句，还有增加、修改和删除数据库内容的语句。除此之外，正如它的名字所表示的，SQL 有查询数据库以提取特殊数据的能力。

SQL 的最初版本是 Sequel，它是由 IBM 公司在 20 世纪 70 年代开发的。1986 年，美国国家标准协会颁布了 SQL 标准，作为商业数据库语言访问关系型数据库的基础。

我们来举一些简单的查询。select 语句是达到这个目的的主要工具，基本的 select 语句包括一个 select 短语、一个 from 短语、一个 where 短语：

```
select 属性 from 表名 where 条件
```

select 短语决定返回哪个属性。from 短语决定对哪个表进行查询。where 短语对返回的数据加以限制。例如：

```
select 标题 from 电影 where 分级 = 'PG'
```

这个查询的结果是列出“电影”表中所有 PG 级别的电影标题。如果没有特别要求可以去掉 where 短语：

```
select 名字，地址 from 客户
```

这个查询返回“客户”表中所有客户的姓名和地址。

这里只列出几个 select 语句的例子，SQL 支持许多不同形式的 select 语句。同样，SQL 的插入、更新和删除语句也使用户可以改变表中的数据。

4.4 数据库和网络

随着网络的发展，网络服务也同样有了很大的发展。许多新型的服务就是有数据库支持的网站。数据库在万维网上的使用极为普遍。事实上，几乎所有通过网站提供产品或公司信息、在线订购或类似服务的公司都在用数据库。最常见的应用包括客户端-服务器数据库处理，用户的浏览器就是客户端软件。

数据库的应用有很多年的历史了，在万维网出现之前很长时间内数据库应用就已经在用网络技术了。银行出纳员所使用的点服务系统就是很明显的早期网络数据库应用的例子。终端装在银行的分支机构里，对银行中央数据库的访问通过一个广域网提供。这些早期的应用只局限于一些能够支付得起这种专门的终端设备的机构和那些能够建设和拥有网络设施的机构。

万维网提供了便宜的、无所不在的网络系统。它有一个标准化的网络浏览器软件用户库，运行于各种各样普通计算机上。对于开发人员来说，可以对用户的文件和程序请求做出回答的网络

服务器软件随手可得，而且已经有几种脚本语言可以用来开发用于网络服务器和网络协议的程序。

网络数据库使用示例

- 获取信息

从本质上来说，数据库就可以提供网络上的信息获取。实际上，它就是个等待提取的存放数据的大仓库。数据保存在数据库中，网站访问者可以发出请求浏览这些内容。这些资料可以是产品信息、网页、出版物、地图、照片、文档等。

- 电子商务

数据库在网络上的另一广泛应用是支持和推动电子商务。目录信息、定价、客户资料、购物车目录以及其他信息都可以保存在数据库中，用合适的脚本或程序把它和网站链接起来并按需要存取。

- 动态网页

动态网页是按照搜索的结果或者用户输入的一些其他请求返回给客户所需要的内容（如文本、照片、表格等）的网页，又称 dynamic HTML（动态网页）或者 dynamic content（动态网页）。"动态"这个词与网站用在一起是指每个用户所得到的网页是个性化的，而不是一成不变的静态网页。

网络数据库是如何工作的

大多数网络数据库的应用是通过三个逻辑应用层把网页和数据库连在一起。底层是数据库管理系统（DBMS）和数据库。顶层是客户网页浏览器作为接口。位于这两层中间的是称作中间件的软件，它通常是用网络服务器脚本语言开发的，它可以与 DBMS 交互，也可以编译并产生网页显示在客户的网络浏览器上。

如图 4-4-1 所示，通常由用户首先发出请求，对网络数据库提取信息或存入数据。常见的请求方式有填写网页表单，选择网页显示的菜单选项和单击屏幕上的广告。网络服务器收到请求后，将其转换成数据库查询程序，并通过中间软件传送到数据库服务器。数据库服务器提取合适的数据再通过中间件反馈给网络服务器，作为网页显示在用户显示屏上。用于与数据库和 Web 网页交互的最常用的中间件是 CGI 和 API 脚本程序。PHP 和 ASP 是更新的脚本程序，现在正越来越流行。

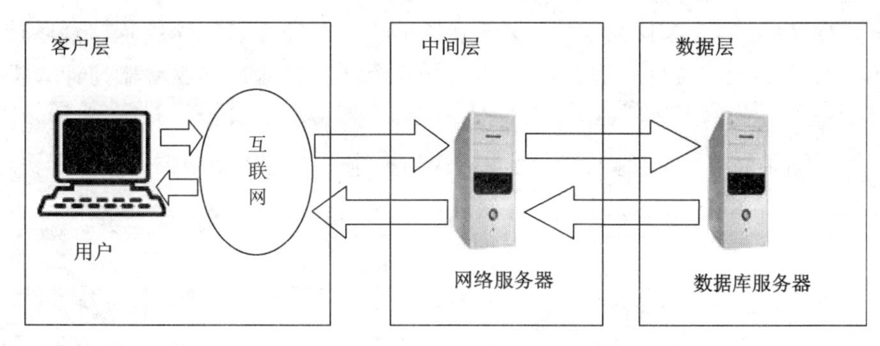

图 4-4-1 网络数据库的工作原理

第5章 多 媒 体

5.1 多媒体简介

什么是多媒体？

回答这个问题并不容易。迄今为止，对多媒体的定义有好几种。就字面意义而言，“多媒体”这个词来自multi和media的合成，其核心是媒体。实际上，多媒体是计算机与视频技术的结合。这意味着计算机信息可以通过传统媒体（文本、图画等）以外的音频、图像、影像、视频和动画等形式表现出来。多媒体之所以能够实现是依靠数字技术。多媒体代表数字控制和数字媒体的汇合，个人计算机是数字控制系统，而数字媒体是当今音频和视频最先进的存储和传播形式。事实上，有人认为多媒体仅是个人计算机和电视的结合。当个人计算机的能力达到实时处理电视和声音数据流的水平时，多媒体就诞生了。

多媒体能做什么？

多谋体可展示信息、交流思想和抒发情感。它可让人看到、听到和理解其他人的思想。也就是说，它可是一种交流的方式。多谋体可帮助人们交流思想、市场策略、公司目标、教育或各种信息，除此以外，还是一种娱乐平台。利用多媒体，可以方便地享受美妙的音乐和参与互动性游戏。

总之，有了多媒体，人们不再是被动的观众，而是可以控制，可以交互活动，可以按需去操作多媒体。也就是说，可以根据自己的需求制作简报，添加文本、图片、视频剪辑或其他自己喜欢的东西。这就是多媒体的力量，也是与传统媒体（如书本和电视）的区别所在。

多媒体配置要求？

多媒体的基本安装配置要求实际上是当今大多数计算机上的标准配置。

第一，多媒体要求具有声音和图形功能，因此，需要有声卡和图形卡。声卡和图形卡都安装在计算机的系统单元内。声卡在机箱背面提供外部接口，可以把扬声器、耳机和传声器等插到该接口。图形卡则是提供到显示器数据电缆的接口。

第二，无论开发还是应用阶段，多媒体产品都需要大量的存储空间。因此，就需要有CD-ROM或DVD驱动器。CD（光盘）是多媒体的主要存储和交换媒体。没有这种方便的光盘，很多多媒体产品无法存储，也就没办法发行和流通。光盘只读驱动器使计算机能够访问音频和软件光盘。

第三，多媒体需要快速的处理器来处理声音和视频所需的大量数据。安装了快速处理器，计算机就能输出更加流畅的视频流，并且保证声音和动作同步。因此，高速处理器芯片是不可缺少的。

其他的多媒体设备还可能包括捕捉设备，如摄像机、录像机、图形采集卡等，交流网络如内网和万维网以及显示设备如高清电视、彩色打印机等。

5.2 多媒体应用

多媒体应用在各个领域包括：广告、娱乐、艺术、教育、工程、医学、数学、商业、科学研究和空间应用，当然不仅仅限于这些领域。几个例子如下：

教育与培训中的多媒体

由于多媒体的引入，学习理论在过去十年间发生了翻天覆地的变化。多媒体在学校教育中发挥了巨大作用。对老师而言，它是一种强大的教学辅助工具。与传统的粉笔加黑板的教学方式相比，PPT 课件更加形象生动，更能激发学生的兴趣。而且，因为多媒体集图、文、声、像、形等于一体，它有助于增强学生的记忆力。对学生而言，多媒体在自学方面具有巨大的潜力。利用多媒体，学生可以获得使他们更清晰理解主题的信息，而不用再冥思苦想课本上抽象的内容。多媒体在学校中的其他用途包括远程教育、校园网络、基于计算机的培训等。

商业中的多媒体

多媒体是商业中的强大工具。它的应用主要包括培训、展示、市场、营销广告、产品演示、手册及网络通信。

基于计算机的培训也被很多公司采用，以培训他们的在职员工和新员工。他们发现这种方法不仅更加形象直观和高效，而且还能节省开支。

利用多媒体，很多公司能够制作出十分漂亮的简报向商业人员介绍信息。

与传统的电视广告和报纸广告相比，多媒体制作光盘提供了一种全新的广告形式。它更有趣，更容易让人接受。商家可以利用这些多媒体演示光盘将自己的产品表现得淋漓尽致。

视频会议（图 5-2-1）是网络通信的一个典型例证。它使得在不同地点的若干与会者能够通过计算机网络传递音频和视频信息，以此来举行会议。

图 5-2-1　视频会议

娱乐中的多媒体

娱乐业广泛使用多媒体。当你想听音乐时，MP3 和 MP4 是不错的选择；当你想看电影时，去电影院或直接用手机看即可；当你和朋友去 KTV 唱歌时，你肯定会用到点歌机（图 5-2-2）；当你想玩游戏时，多媒体的重要性更加明显，因为最早的多媒体程序之一就是为游戏设计的。有了先进的多媒体技术，就能身临其境般地享受那些互动性游戏（图 5-2-3）。如今，流行的娱乐用多媒体还包括 iPhone, iPod touch（图 5-2-4），iPad 等。

图 5-2-2 点歌机

图 5-2-3 掌上游戏机（PSP）

图 5-2-4 iPod touch

公共场所的多媒体

人们常说这是一个信息时代。对很多人来说，随时随地获取最新的信息非常重要。多媒体就为获取信息提供了高效的途径。无论人们在乘坐公车、地铁、出租车还是自己开车，移动电视（图 5-2-5 和图 5-2-6）或其他多媒体设备都可提供大量的信息。甚至走在路上都能通过户外大屏幕（图 5-2-7）获得信息。如果新到一个城市，不必担心，路边的多媒体终端机会帮你指路。在像宾馆、火车站、购物中心和超市这样的公共场所，多媒体通过终端机或服务厅为人们提供信息和服务。作为一种交流信息的有效途径，多媒体将被用于更多的公共场所。

图 5-2-5 地铁移动电视

图 5-2-6 公交车移动电视

图 5-2-7 大型电子显示器

5.3 多媒体工具

要创建多媒体项目，一套基本的工具应包括一个或多个创作系统和众多的文本、图像、声音和视频等编辑软件。其他一些用于截取屏幕图像、转换文件格式以及在合作小组内部传输文件的软件也十分有用。使用这些多媒体工具，你的创意生活将变得更简单。

PowerPoint

作为 Office 套件软件之一，PowerPoint（图 5-3-1）提供了多媒体演示的手段。它的强大功能在于制作幻灯片。用户可以非常容易地在幻灯片上输入标题和文本，并添加图片、表格和图形。如果愿意，还可以改变幻灯片的布局、调整幻灯片的顺序、删除或复制幻灯片。PowerPoint 广泛用于学校中制作课件和商业上制作简报。实际上，它的用途远远不止这些。目前，流行的版本是 PowerPoint 2007、PowerPoint 2010 和 PowerPoint 2016。

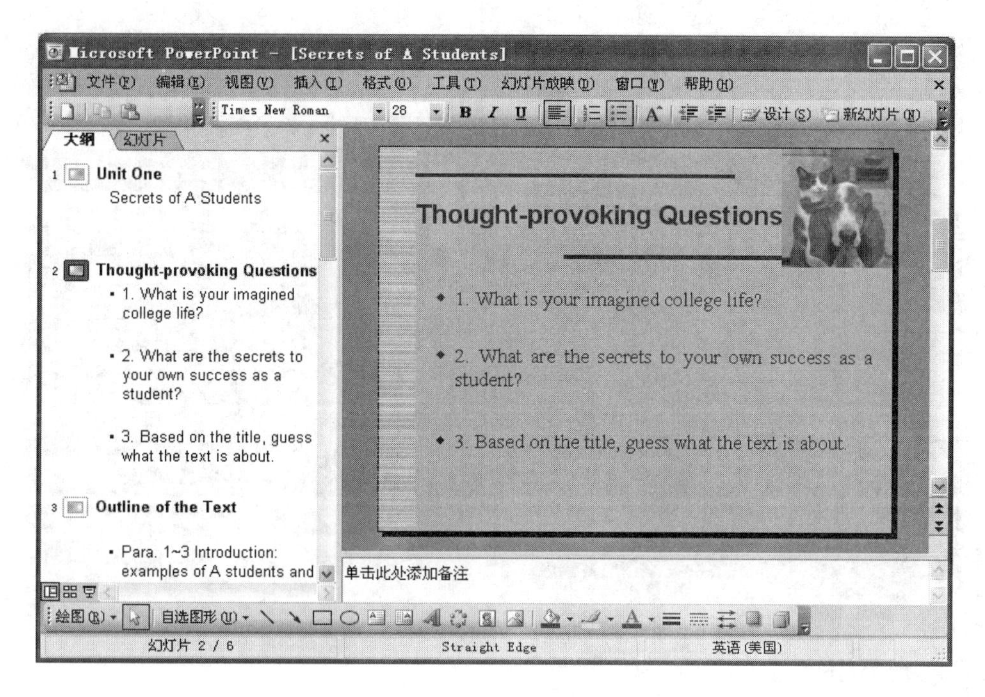

图 5-3-1 PowerPoint 窗口

画图和绘图工具

流行的画图和绘图工具包括 CorelDraw、Illustrator、FreeHand、AutoCAD、Adobe Photoshop、Micrografx Picture Publisher 等。它们大致可以分为两类：矢量绘图软件和位图绘图软件。简而言之，位图和矢量的区别在于位图图像比矢量图像更加真实，但是它比矢量图像需要更大的存储空间。CorelDraw 就是一种矢量绘图软件。它在平面广告设计、商标设计、写意、商品包装设计、漫画创作等领域都很受欢迎。目前的常用版本有 CorelDraw 通用版、CorelDraw X4 和 X5，最新版本是 X6。另一种专业的矢量图形软件是 Adobe Illustrator。它深受平面设计师、专业插画家、生产多媒体图像的艺术家、网页设计师等的青睐。然而，随着软件行业的发展，某一种绘图软件的功能将会不仅仅限于处理一种类型的图像。

三维建模和动画工具

使用三维建模软件，用透视法对物体进行渲染可使物体更加逼真。只要为最终渲染的图像选择正确的光线和透视法，即可创造绝妙的场景。典型的建模软件是 3D Studio Max，通常简称为 3ds Max 或 MAX（图 5-3-2）。它是 Autodesk 公司开发的一种强大的三维动画渲染和制作软件。目前被广泛应用于广告、电影、游戏开发、建筑设计和教育等领域。它的两个最新版分别是为游戏专业人员设计的 Autodesk 3ds Max 2012 和为设计师们设计的 3ds Max Design 2012。

一些其他的三维动画渲染和制作软件还有 Avid Softimage XSI、Sumatra、Alias/Wavefront MAYA、Houdini、LightWave 3D、Animatek World Builder2C、Bryce 和 Poser。

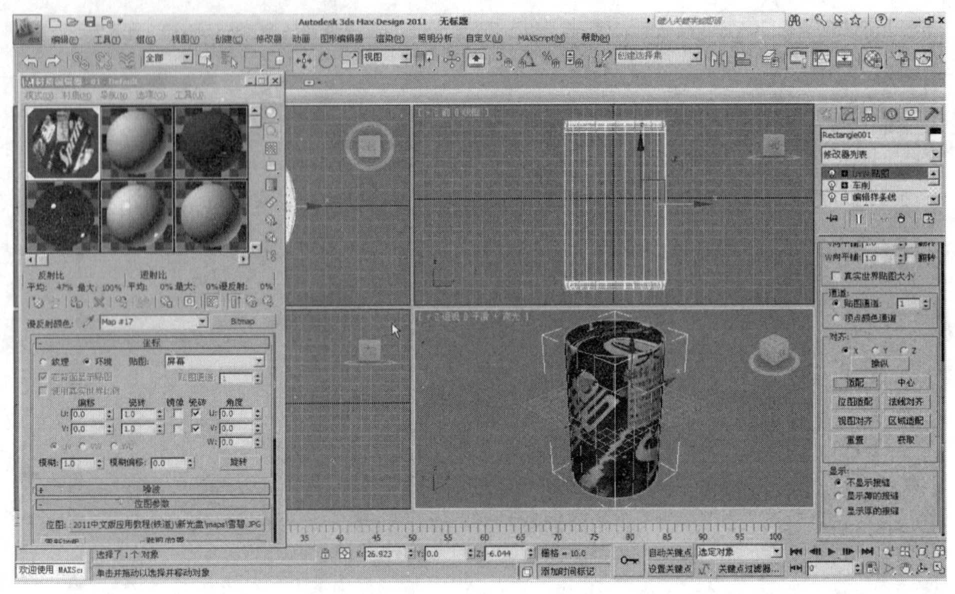

图 5-3-2　3ds Max 窗口

视频和数字电影工具

要从视频制作出电影，需要特殊的电影制作工具。Adobe Premiere 是一种非线性数码影视编辑软件（图 5-3-3）。利用它，能够很容易地对录像机拍摄的视频片段、其他数字化电影片段、动画、扫描图像、数字化音频或 MIDI 文件进行编辑和组合。这些功能也可以由 MediaStudio Pro 或 After Effects 来完成，它们也是电影制作工具。其他影视制作软件还有会声会影、Avid Xpress Pro HD、Sony Vegas 6.0、Digital Fusion 5.0、Discreet inferno 系统等。

图 5-3-3　Adobe Premiere 窗口

第 6 章　计算机网络

6.1　网 络 简 介

计算机网络就是把计算机连在一起，形成一个使用户能共享硬件、软件和数据，并使它们间相互通信的网络。各种网络一般都使用网络服务器用以管理网络资源。网络服务器控制共享打印机和其他硬件以及共享程序和数据的使用。连接因特网，在网络中提供网络信息访问功能的服务器称为 Web 服务器，主要提供网页服务。

网络的分类

计算机网络有多种规模和类型，它可以仅包括几台计算机、打印机和一些其他设备，也可以由分布在广大地理区域上的许多大大小小的计算机组成。网络的一些特性包括它的物理结构、网络规模的大小以及网络间距等。

- 根据规模大小

根据规模或地域范围大小，网络可被分为三种类型：局域网、城域网、广域网。

- 根据连接方法

计算机网络也可以根据用以连接网络中单个设备的硬件技术来归类，这种设备有光纤、以太网、无线局域网、电力线通信。以太网使用物理线路来连接设备。采用的设备通常是集线器、交换机、网桥和/或路由器。无线局域网技术不用布线就能连接设备。这些装置使用无线电波作为传输介质。

- 根据网络拓扑结构

计算机网络可根据网络的拓扑结构归类，分为总线型网络、星状网络、环状网络、树状拓扑网络等。

下面根据网络的规模大小列出最常见的计算机网络类型。

➢ 个人区域网（PAN）

个人区域网（PAN）是用于连接人们近距离使用的计算机设备的一种计算机网络。用于个人区域网的设备通常包括打印机、传真机、电话、扫描仪等。个人区域网涉及范围通常约 20～30 英尺（约 6～9 米）。个人区域网络，可有线连接计算机总线。使用红外线和蓝牙等网络技术，可实现个人区域网络无线化。

➢ 局域网（LAN）

局域网是一种高速网络，覆盖一个较小的地理区域，例如一个家、一间办公室或一栋大楼。它通常连接工作站、个人电脑、打印机、服务器和其他设备。目前的局域网大多基于以太网技术。

➢ 校园区域网络（CAN）

校园区域网络连接两个或两个以上的局域网，局限于大学校园或者军事基地这样具体明确并且邻接的地理区域内。

校园区域网络可被视为城域网的一种，但一般仅限于比典型的城域网小的区域范围内。这个词最常用来讨论邻接地区的网络实现。局域网连接的是相对短距离的网络设备。网络中的办公楼、学校或家庭通常包含一个单一的局域网。

➢ 城域网（MAN）

城域网连接两个或两个以上的局域网或校园区域网，将一个城市中各办公大楼的网络连接起来，但仅限于这个城市范围内。

➢ 广域网（WAN）

广域网涵盖了较广的地理区域（一个城市到另一个城市，一个国家到另一个国家），经常使用共同运营商如电话公司所提供的传输设施。最广泛使用的一种广域网是互联网。广域网技术实现的功能一般为OSI参考模型的最低三层：物理层、数据链路层、网络层。

➢ 互联网络

任何公共、私人、商业、工业或政府网络之间或之内互连的网络也可能被界定为一个互联网络。互联网络使用互联网协议。至少有三种互联网络：内联网、外联网、互联网，这要看谁管理，谁加入了网络。内联网和外联网可能会或不会连接到互联网。如果内联网或外联网连接到互联网，通常是未经适当授权，禁止被互联网访问。

内联网是一套相互关联的网络，处于一个单独的行政机构的控制下，使用互联网网络协议，并使用基于IP的工具，如Web浏览器和FTP工具。行政机构的内联网不对外界所用，只允许特定用户使用。内联网通常都是公司为其员工建立的内部网络。大型的内联网通常会有自己的Web服务器，便于向用户提供浏览信息。

外联网就是在有限的范围内，连接一个单一的组织或机构，但这个组织或机构也与其他一个或多个不一定是值得信赖的组织或机构进行有限的连接。例如，公司的客户可进入内联网的一部分，此种方式就建立了一个外联网，而同时从安全的角度，客户可能不会被视为"值得信任的"。虽然根据定义，外联网不能构成一个单一的局域网，它必须与外部网络至少有一个连接。但在技术上，外联网也可被归类为一个局域网、城域网、广域网或其他类型的网络。

➢ 因特网

因特网，由全球联网的政府的、学术的、公共的和私人的网络构成，建立在美国国防部高级规划署开发的网络基础上，常用带有大写"I"的"Internet"来区别于其他种类的互联网络。因特网用户使用通过地址注册表获得的一系列因特网协议和IP地址。服务提供商和大企业通过边际网关协议，在他们可使用的地址范围内交流信息。

基本硬件组件

所有网络都由基本的硬件部件组成，这些部件把网络结点，如网络适配器（NIC）、网桥、集线器、交换机、路由器等互连起来。

- 网络适配器

网络适配器（图6-1-1），也称网卡，是一块用来通过计算机网络进行沟通的硬件设备。它提供了一个访问网络介质的物理渠道，往往通过使用媒体访问控制子层协议地址，提供一个低级别的地址系统。用户利用电缆或无线方式进行相互连接。

- 转发器

转发器（图6-1-2）是网络中放大信号的设备，能够接收信号并能在更高层次上或在受到阻塞的一边重新发射信号，使信号可以无衰减地覆盖更长的距离。在大多数双绞线以太网结构中，电缆运行长度超过 100 m 就需要转发器了。

图 6-1-1　网卡

图 6-1-2　转发器

• 集线器

一个集线器（图 6-1-3）包含多个端口。当一个数据包到达一个端口，就被复制到集线器的所有端口进行传输。当数据包被复制，框架里的目的地址在传播地址中并没有改变。它以一种简单的方式，复制了连接到集线器的所有结点的数据。

• 网桥

网桥（图 6-1-4）在 OSI 模型的数据链路层（第 2 层）连接多个网段。网桥不像集线器那样，不加区别地把经过的通信量复制到所有端口，但能得知哪个 MAC 地址是可以通过特定的端口到达的。网桥一旦联系到一个端口和一个地址，它将会把那个地址的通信量只传送给那个端口。

图 6-1-3　集线器

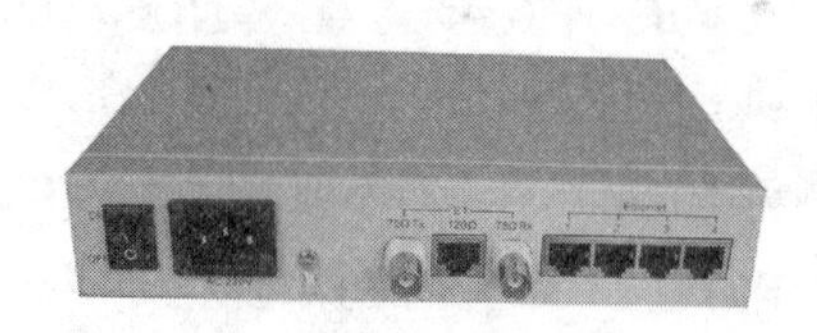

图 6-1-4　网桥

• 网络交换机

网络交换机（图 6-1-5）是一种执行交换的装置。具体来说，它转发和过滤连接电缆的端口之间的 OSI 模型第二层的大块数据通信，这些端口建立在数据包中 Mac 地址的基础上。网络交换机有别于集线器，因为它只向涉及通信中的端口转发大块的数据通信，而不是向连接的所有端口转发。严格说来，网络交换机不能路由建立在 IP 地址（第 3 层）基础上的通信量，而 IP 地址对于网段之间通信或一个大型或复杂的局域网之间的通信则是必要的。出于大部分或全部的网络都能直接与其他的交换机相连的目的，一个交换机通常有众多的端口。交换机可运行在 OSI 模型的一个或多个层次，包括物理层、数据链路层、网络层或运输层。同时在一个以上的这些层上运作的设备称为多层交换机。

• 路由器

路由器（图 6-1-6）是转发网络之间数据包的网络设备。它能确定最佳数据包转发路径。路由器工作在网络层的 TCP / IP 模式或 OSI 模型的第 3 层。路由器还提供相似或不同的媒体文档间的互连。网络上的路由器可共享网络信息，它们能够选择两个子网之间的最佳路线。如果网络一部分服务拥挤或无法服务，一个路由器可选择一条路径来传送数据包。路由器至少与两个网络相连，通常连接两个局域网或广域网，或连接一个局域网及其 ISP 的网络。为家庭甚至办公室使用的 DSL 和电缆调制解调器，已集成于路由器，允许多个家庭/办公室的计算机通过相同的连线访问互联网。许多这样的新设备还包括无线接入点或无线路由器，以便 IEEE 802.11b/g 无线功能设备能够连接到网络，而不需要电缆连接。

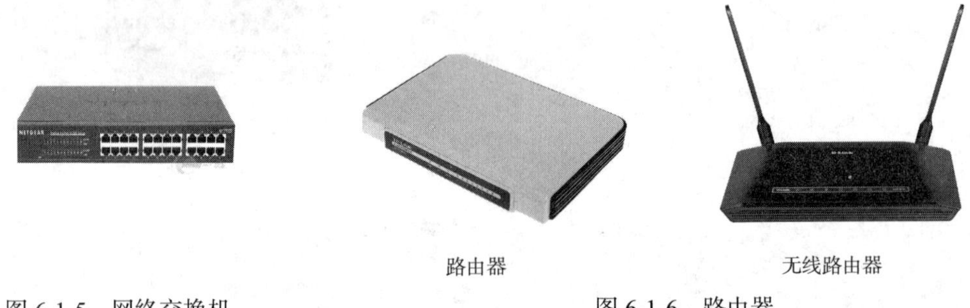

图 6-1-5　网络交换机　　图 6-1-6　路由器

网络协议

网络协议为网络设备间的通信界定了规则和惯例。计算机网络协议普遍使用分组交换技术以数据包的形式来传送和接收信息。网络协议含有一系列机制，使设备能够识别和相互连接，并形成一套规则，来详细说明数据如何分组成不同的信息被发送和接收。一些协议也支持邮件确认和为可靠的和/或高品质的网络通信设计的数据压缩。已开发数百种不同的计算机网络协议适用于特定用途和特殊环境。

在一个通信连接中，协议在各层都存在，一台设备上的第 *n* 层与另一台设备上的第 *n* 层进行通信的规则就是第 *n* 层协议。常见的互联网协议包括传输控制协议（TCP），互联网协议（IP），超文本传送协议（HTTP）。传输控制协议（TCP）在信息数据包传输层，使用一套规则用以规范互联网站点之间的信息交换；互联网协议（IP）在互联网地址层，使用一套规则用以规范互联网站点之间的信息的发送和接收；而超文本传送协议（HTTP）是用于传送任何出现于万维网上的数据的协议。

6.2　局　域　网

局域网是指覆盖在相关小范围内的计算机网络。局域网最初是在大学和政府的研究机构发展起来的。大多数局域网被限制在一幢或一群建筑物当中使用。但是各个局域网可通过电话线与无线电波连接，来进行科学数据的交换以及计算处理能力的共享。

局域网给计算机用户提供许多优势，如共享访问设备和应用软件，连网的用户之间交换文件，通过电子邮件和其他软件在用户间通信。例如，一个图书馆可为用户提供一个有线或无线局域网来互连本地设备，如打印机和服务器，也可连接到互联网。对于一个有线局域网，如图书馆的个人计算机，通常由电缆连接，通过互连的设备系统，运行 IEEE 802.3 协议，并最终连接到互联网。无线局域网使用不同的 IEEE 协议，802.11b 或 802.11g。工作人员可以使用彩色打印机，访问学术网络和因特网。每个工作组可以使用其本地打印机。但请注意，工作组以外的用户不可以使用本打印机。

与广域网（WAN）相比，局域网有其鲜明特点，它有较高的数据传输速率，使用地理范围较小，不需要租用通信线路等。目前的以太网或其他 IEEE 802.3 局域网技术运行速度高达 10 Gbit/s。

- 局域网设备及其组织方式

局域网设备通常包括中继器、集线器、局域网扩展部件、网桥、局域网交换机和路由器。中继器用于互连媒体段扩展网络。中继器把一系列的电缆段当成一个单一的电缆段。中继器从一个网段中接收信号、放大、延迟并重发这些信号到另外的网段，通过这种方式避免信号在长距离电缆和大量连接设备中减弱。集线器用于连接多用户站点。交换机基本上是多端口网桥，工作时把网络分成很多段，每一段的本地业务运行都不会对其他网段产生干扰。交换机记住了常驻在每个端口的那些地址，并对数据进行相应的交换。

每一个局域网其设备都有它自己的几何分布，局域网拓扑结构定义了网络设备的这种组织方式。在局域网中，有三种基本的拓扑结构：总线型结构、环状结构和星状结构。下面各图显示了其结构形状。

在总线型拓扑结构中（图 6-2-1），所有设备都连接到一根中心电缆或中枢上。总线型拓扑结构相对来说比较便宜，便于安装较小的网络。在环状拓扑结构中（图 6-2-2），所有设备相互连接成一个闭环，这就便于每一设备都可连接到另外两台设备，一侧一台。环状拓扑结构相对来说比较昂贵，难于安装，但能提供高速带宽并能进行远距离传输。在星状拓扑结构中（图 6-2-3），所有设备都连接到一个中心集线器上。这种结构相对来说易于安装和管理，但由于所有数据都必须通过中心的集线器，容易造成堵塞。

图 6-2-4 显示了树状拓扑结构，它常被用于其他网络。

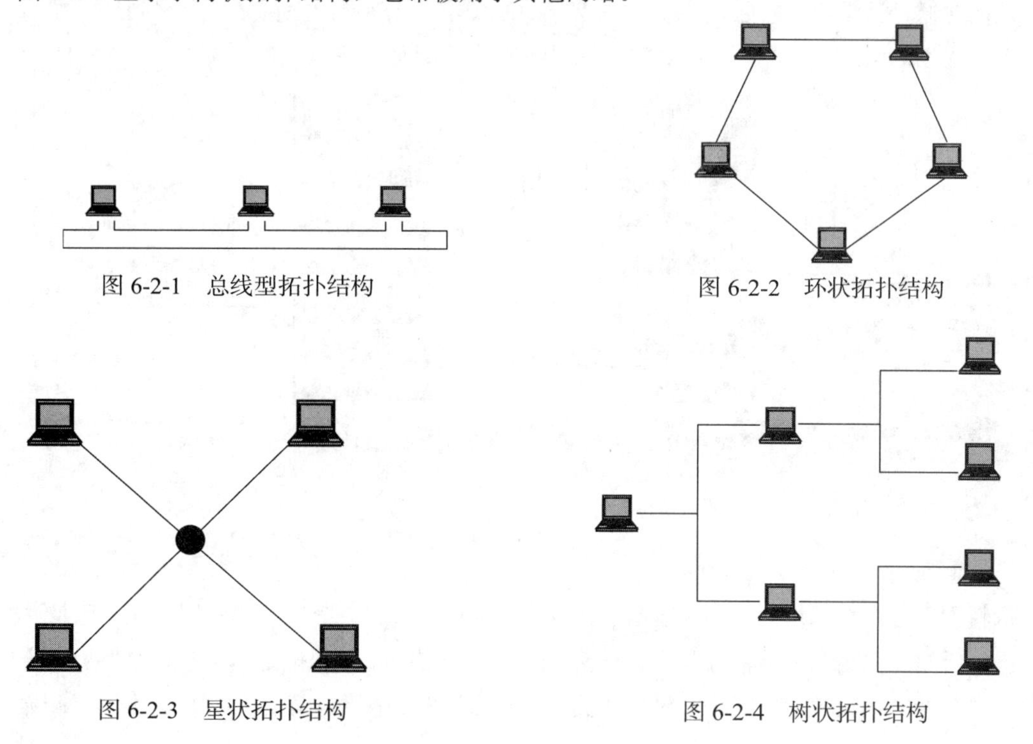

图 6-2-1　总线型拓扑结构

图 6-2-2　环状拓扑结构

图 6-2-3　星状拓扑结构

图 6-2-4　树状拓扑结构

- 关于 ISO OSI

ISO OSI 指的是国际标准化组织/开放系统互连参考模型。它提供了一个通信系统的完整功能模型，是通信方面最重要的系统体系之一。它是一个概念模型，由七层组成，每一层的目的都是为其上层提供某些特定的服务，因此只专注于各层之间如何协调工作。图 6-2-5 展示了开放系统互联参考模型的七层。

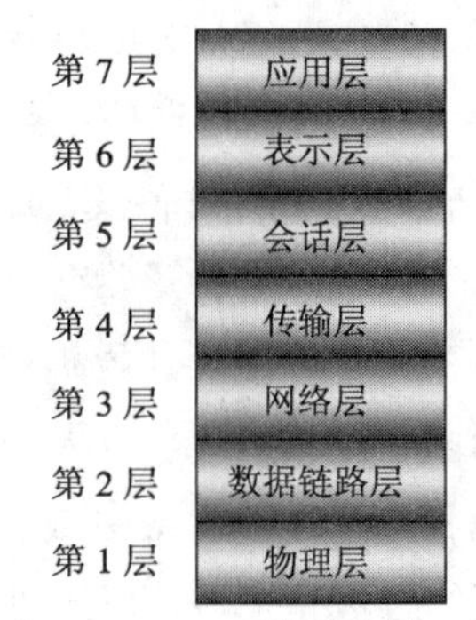

图 6-2-5　OSI 模型中的 7 层

在 OSI 模型中，这七层被分成高层和底层两大类。高层包括应用层，表示层和会话层，定义了应用程序的功能。底层包括传输层、网络层、数据链路层和物理层，处理与数据传输有关的内容。每一层都有它自己的特定的功能。第 1 层也是最底下的一层叫物理层，负责如何在一条通信信道上传送原始位数据。第 2 层叫数据链路层，提供点对点的分组传送并负责在网络介质上传送/接收结构化的位流。第 3 层是网络层，当网络层从源主机接收到信息的时候，它便把这些信息进行数据分组，即变成信息单元，并确保这些信息单元在目的端被正确地接收。传输层的功能是保证数据成功地发送到目标设备。会话层使得不同设备间建立对话并交换数据。表示层关心的是所传送信息的语法和语义并负责对数据进行编码和加密。应用层为应用程序提供服务，以确保网络中的应用程序间进行有效交流。

计算机系统上的一个软件应用向另一个系统上的应用传送出的信息必定要经过 OSI 的各个层。然而，某台计算机上的第 n 层只与另一台计算机上的第 n 层进行对话。确切地说，两个计算机 A、B 之间要传递信息，数据就会从计算机 A 的 OSI 一端向下逐层传递，而在另一端，当信息到达的时候，数据就在接收计算机 B 中沿着 OSI 的各层逐层向上传递，最终到达终端用户。图 6-2-6 形象演示了计算机中 OSI 模型的各层间的信息传递。

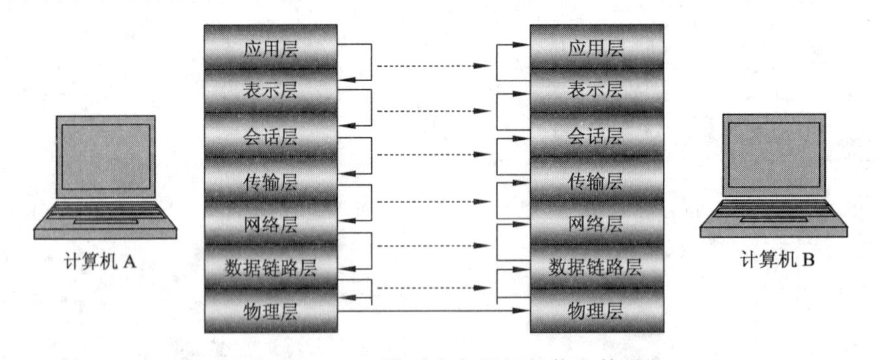

图 6-2-6　OSI 模型中各层间的信息传递

6.3 因　特　网

因特网是一个全球性的网络，连接了数以百万计的计算机。世界各地的用户连入因特网来交换数据、新闻和观点以达到资源共享、互相沟通。互联网的流行称谓是信息高速公路。不论你是否要找最新的财经或政治新闻，还是浏览图书馆目录、发送接收电子邮件，与同事网上聊天，或加入到一个生动的辩论中，互联网都能帮你达到目的。互联网由数以千计的较小的网络组成，因此从技术上说，没有人运行互联网。正是由于许多用户寻求新的方法来创造，显示和检索因特网上的资料，互联网才得以繁荣和发展。

不存在任何一个中央机构向单个的互联网用户收费。相反，使用互联网的个人和机构为他们所分享的服务向当地或地区的互联网服务提供商支付费用。反过来，那些规模较小的互联网服务供应商可能会从一个更大的网络购买服务。因此，基本上说来，利用互联网的人们都以某种方式支付部分费用。要想连接到互联网，用户必须通过服务供应商。根据不同的选择，每月付费就可以了。由于选择的不同，访问时间可能会有所不同。

- Web 客户机/服务器

因特网的主要作用之一就是共享资源。这种共享通过运行在不同计算机上的两种程序来实现。一种程序称为服务器，能够提供一种特别的资源。另一种程序称为客户机，能利用这种资源。确切地说，接收网络服务的工作站称为客户机，而处理网络服务请求的计算机称为服务器。如果有必要连接到另一类型的服务，例如，建立一个远程对话或下载文件，客户机都能做到。客户机与服务器连接，会代表用户索要信息。

所有基本的互联网工具，包括远程登录、FTP、Gopher、万维网，都建立在一个客户端和一个或多个服务器协同工作的基础之上。每种情况下，用户与客户端程序互动，管理数据如何呈现给用户或用户寻找资源的方式等细节问题。 反过来，客户与信息驻留的一个或多个服务器互动。服务器收到请求并运行它，然后发出结果。

- 万维网/因特网

万维网既不是一个网络也不是因特网本身，它是因特网上发展最快的应用软件之一，Web 是客户端和服务器利用因特网进行数据交换的一个系统，由遍及因特网的无数网页组成，可以访问遍布于因特网上的成千上万台计算机中的链接文档。浏览万维网，用户可以获取产品信息、实时新闻、天气预报、航班和列车时刻表、各种出版物、音乐和电影下载等信息，也可以进行购物、储蓄、股票买卖和其他各种在线金融交易活动。如果想从 Web 网站上得到所需网页，必须输入该网页的因特网统一资源定位（URL）命令。URL 指的是统一资源定位器，用以指名 Web 服务器的因特网地址和文件名。Web 服务器要明白 Web 浏览器发出的请求，双方必须采用同一标准协议。在 Web 浏览器和 Web 服务器之间进行通信的标准协议是超文本传输协议，它规定了数据如何通过网络传输。

- IP 地址/域名系统

IP 地址和域名用来标识因特网上可访问的计算机。IP 地址采用数字表示，如 202.112.7.12。在互联网上进行通信的每台计算机都分配有一个 IP 地址。IP 地址用来唯一标识计算机设备，将它与互联网上的其他计算机设备区别开来。

由本地主机和它的域名所构成的域名系统，映射了数字表达的互联网地址和主机名（域名）之间的关系。例如，www.pku.edu.cn 是个域名，而 202.112.7.12 则是其相应的数字式互联网地址。IP 地址很难记忆，而域名使用字母表达，便于记忆。域名包括如下信息：计算机名、组织名、组织类型和所处地理位置。域名系统的最后一个部分如.com, .edu, .gov 等，称为顶级域名，往往指示主机所在机构的类型。如 edu，gov，com，net，org，mil 等分别代表教育、政府、商业、网络、机构、军事等。通常也用最后两个字母表示国家和地区，如 uk 表示英国。当今，域名尤其是最重要的两个组成部分，如 sina.com，已经成为许多公司“品牌”的一个宝贵部分。因此域名的分配已经变得高度政治化。

- 因特网服务

目前，人们可享用到因特网提供的各种服务，包括 FTP、电子邮件、WWW、新闻、网络电视/电影、网络音乐、Internet 联机聊天系统、远程登录和其他服务。许多新用户是从上网、收发电子邮件以及查找信息开始的。例如，电子邮件服务能够可靠地传送和接收信息。每个信息从一台计算机被发送到另一台计算机，最终到达目的地。网络购物也是一种重要的网络服务。网络购物在因特网使用方面发展迅速，并且变得越来越流行。通过访问网站，人们可以从大公司或小零售商直接购买产品。足不出户，人们轻点鼠标就可寻找他们所需要的任何物品，然后网上支付即可。当用户买自己想买的东西时，可以在不同的网站搜寻不同的网店，比较其价格再去预订，然后安排送货方式，东西很快就会送到其手中。整个交易过程几分钟之内就能完成，省得你去店铺购买。因此，网络购物能够省时省钱。然而，网络购物也有一些缺点。最大的缺点就是用户只能看到产品的画面。因此，摸不着这些东西，也不能试用，就会失去这些乐趣。也不能检查货物的质量。网络购物的另一个缺点就是缺乏安全性。网络空间的商业角落里仍存在很多陷阱。许多人担心网络的安全性和可靠性。由于网络购物已变得非常广泛，尽管因特网用户确实有安全性方面的担忧，但这并没有放慢网络购物者数量增加的速度。

6.4 网络安全

随着网络的迅速发展以及新科技的不断引进，人们创造并享受着新的商业机会。但人们又不得不保护他们的计算机以免受黑客攻击。入侵者通过网络搜寻信用卡号码、银行账户信息以及其他任何能找到的东西。入侵者还想要获取你计算机的硬盘空间、高速处理器、因特网连接资源等。而且，他们利用这些资源攻击互联网上的其他计算机。因此，网络安全问题越来越受到人们的关注，成为计算机网络一个必不可少的部分。计算机网络防御设施里所做出的各种规定、网络管理员为保护网络和网络访问资源免遭未经授权的访问的策略以及这些措施有效的结合，便构成了网络安全。不同的情况下，网络安全管理是不同的。家或办公室仅需要基本的网络安全，而大型企业将需要高昂的维护和先进的软/硬件，以防止黑客和垃圾邮件的恶意攻击。要使计算机安全并不是一件轻松的工作，下面介绍与网络安全紧密相关的一些问题。

- 数据加密和密码

最安全的措施包括数据加密和密码。数据加密是把数据翻译成一种难解的形式，如果没有破码机制就无法阅读它。只有那些被授权的人们或系统才能解读。密码是一个让用户访问某个特定程序或系统的秘密的词或短语。像房门钥匙一样，每台计算机的密码应该是唯一的。每台计算机和每种服务，都应该使用密码。

- 防火墙

防火墙是一个由硬件或者软件组成的系统，用以阻止对私人网络未经授权的访问，保护计算机或者计算机网路免受攻击。系统管理员往往把代理防火墙与包过滤防火墙结合在一起，以创造一个高度安全的系统。大多数家庭用户使用软件防火墙。这种类型的防火墙可创建日志文件，记录所有与个人计算机连接的细节，包括尝试连接。

• 反病毒软件

有些人或公司恶意编写一些计算机程序，如病毒，蠕虫，特洛伊木马和间谍软件等。这些程序都被定为不想要的软件，但它们自己通过欺骗的方式安装在用户的计算机上。特洛伊木马程序掩盖其真正的目的或包含了用户不想要的隐藏功能。当特洛伊木马病毒运行时，它就会对计算机系统做坏事，而看起来却像在做对系统有益的事情。蠕虫病毒的特点是有能力自我复制，除了通过向第三方软件加入自己的代码之外，病毒都是相似的。病毒或蠕虫病毒一旦感染了一台计算机，它通常会感染其他程序和其他计算机。蠕虫病毒可能在一台计算机中不停复制自己，最后导致计算机崩溃。病毒也能使系统性能降低并引起奇怪的系统行为。在许多情况下，无论是有意、无意或恶意破坏，病毒都会对计算机造成严重危害。为防止病毒损害，用户通常安装防病毒软件，它们在计算机后台运行，检测任何可疑的软件并防止它运行。

• 蜜罐技术

蜜罐技术的关键作用是引诱网络访问资源，作为监测和早期预警工具在网络里使用。那些企图向这些诱饵资源妥协的攻击者所用的技术，在攻击期间和之后被研究，以留意新的开发技术。这些分析可用来进一步加强被蜜罐技术保护的实际的网络安全。

• 反间谍软件

安装反间谍软件是保护计算机的一种有效方式。间谍软件是一种不经过用户明确许可就能在计算机上运行的软件。它常常从用户的计算机上收集私人信息，并将此数据通过互联网发送给软件制造商。广告软件很像间谍软件，未经同意就能运行在用户计算机上。然而，它不取走任何信息，通常在后台运行，只是随机显示或有针对性的弹出广告。在许多情况下，这也会使计算机速度减慢并可能导致软件冲突。

目前，新兴的云安全技术可以实时确认并阻断各种威胁，不让它们到达用户计算机，成为强大的防护武器，保护着互联网和网络用户。

附录C 科技英语的翻译技巧及语法

科技英语的翻译技巧

要掌握科技英语的翻译技巧，首先应该了解科技英语的词汇和句式特点。词汇方面，科技英语的基本词汇与普通英语无异，只是出现了大量专业技术词汇和术语，这些专业技术词汇在词形上可能是某专业特有的，也可能与普通英语词汇相同，只是词义不同；句式方面，鉴于科技语言的客观性、严谨性和精练性，科技英语句式中多使用被动语态、非谓语动词、后置定语和名词化结构。

针对科技英语的语言特点，在从事翻译的过程中通常采用的一般技巧如下：

（1）选择恰当的词义

由于英语词汇来源复杂，同一单词在不同专业中词义可能相差很大，词性也可能出现变化。例如，key 通常指钥匙，但在计算机英语中指键盘上的键，在音乐中指调。又如 power 在数学中指乘方，在光学中指放大倍数（率），在电学中指电力供应。所以，在翻译科技文章时一定要注意选择恰当的词义。

（2）适当增减词汇

翻译过程并不要求两种语言的句子在词汇数量上一定要相等，可以根据实际情况适当增加、减少或重复一些词。例如：

① The rise of the world wide web has given embedded designers another quite different opinion.

万维网的出现为嵌入式系统的设计者们提供了一个全新的选择。（增加了“系统”一词）

② Electronic commerce enables new forms of business, as well as new ways of doing business.

电子商务使新型商务及新的商业模式成为可能。(doing 并未译出)

（3）转译词性或句子成分

科技英语中大量地使用名词化结构，而汉语习惯使用动词，所以在翻译时可以进行必要的词性转换；而词性的转换有时又会引起句子成分的转换。例如：

① A speedy processor chip plus a CD-ROM drive or DVD drive are desirable.

需要高速处理器芯片和 CD-ROM 驱动器或 DVD 驱动器。

（形容词转译为动词，句子成分也由英语句子中的表语转换成了汉语中的谓语。）

② For high volume systems such as portable music players or mobile phones, minimizing cost is usually the primary design consideration.

对于便携式音乐播放器或手机这类大容量的系统，成本最小化通常是设计时考虑的主要因素。

（两个名词都转成了动词，句子成分也发生了转变。）

下面根据科技英语中使用最多的句式，具体介绍被动语态及定语从句的翻译方法。

1. 被动语态的翻译方法

（1）译成汉语被动句

科技英语中的被动句在汉语中大多被翻译为主动句，少数也可以译为汉语的被动句。当译成汉语的被动句时，可以在译文中加入“被”“由”“使”“为……所”等表示被动。例如：

① These are called removable-media storage systems.

这些被称为可移动的存储系统。

② Size is indicated by the diagonal length of the monitor's viewing area.

尺寸由显示器对角线的长度表示。

（2）译成汉语主动句

当翻译成汉语主动句时，可采取的具体方法有：

① 原句中主语不变，通过在译文中加词（如经过、加以等）表达被动含义。例如：

The components of computer hardware are listed in Table 1-1 and described here.

计算机硬件的组成列在表 1-1 中，并且加以解释。

② 当被动句主语为形式主语 it 时，可以用泛指词（如人们、大家、我们）作主语，而把英文句子中真正的主语译为宾语。例如：

It can be seen that these instructions are for computer operations.

大家可以看到这些指令是用来操作计算机的。

③ 翻译成汉语的无主句，即省略主语。例如：

So a sound card and a graphics card are needed.

因此，需要有声卡和图形卡。

2. 定语从句的翻译方法

（1）合译法

该方法适用于简单结构的定语从句，尤其是限制性定语从句。其译法是将从句翻译为“……的”，置于被修饰的名词前。例如：

① Software refers to programs that control the operation of the hardware.

软件是指控制硬件操作的程序。

② Here we will introduce the devices which enable people and computer to communicate.

以下将介绍能使人机交流的设备。

（2）分译法

该方法适用于非限制性定语从句和长的限制性定语从句。其译法是将从句翻译成独立的句子，以关系代词所指代的名词、代词或者“这”作为其主语；该从句译文置于主句译文之后。例如：

① Music can be input into a PC through MP3 players or MP4 players, which are also output devices.

音乐可以通过 MP3 或 MP4 播放器输入个人计算机，MP3 和 MP4 播放器同时又是输出设备。

② The motherboard is located at the bottom of the desktop case or the side of a tower case, on which there are several sockets.

主板位于台式机箱的底部或塔式机箱的侧面，它上面有很多凹槽。

③ Linux is a clone of the UNIX operating system that runs on Intel 80x86-based machines, where x is 3 or higher.

Linux 是 UNIX 操作系统的克隆，它运行在基于 Intel 80x86 的计算机上，这里的 x 指 3 或者大于 3 的数字。

（3）转译法

该方法适用于某些特定的定语从句。

① 当含有定语从句的主句句型为 there be 时，从句谓语译为主句的谓语。例如：

There are four separate products that form the Windows 2000 line of operating systems：…

四个独立的产品形成了 Windows 2000 操作系统系列：……

② 当定语从句起状语的作用时，可以将定语从句转译为状语。例如：

Computers, which have many advantages, can not carry out creative work and replace man.

尽管计算机有许多优点，但不能进行创造性工作，也不能代替人。

（4）注释法

该方法适用于插在句子中起补充说明作用的非限制性定语从句，译法是将其放在括号内注释。例如：

In contrast, “storage” refers to the permanent storage available to a PC, which is also called secondary storage—usually in forms of the PC’s hard drive, floppy disks，CDs, etc .

相比而言，外存（又称辅助存储器）是指以硬盘和 CD 等形式在个人计算机上使用的永久性存储器。

练　　习

将下列句子翻译成中文，并指出使用了以上何种翻译方法

1．The input/output devices are sometimes called peripheral devices.

译文：____。

翻译方法：________________________。

2．These three sections—the arithmetic logic section, the control section, and the memory section—are found in CPUs of all sizes.

译文：____。

翻译方法：______________________。

3．A computer is a fast and accurate system that is organized to accept, store and process data, and produce results under the direction of a stored program.

译文：____。

翻译方法：______________________。

4．This is a major reason why personal computers are so popular today.

译文：____。

5．Java is an object-oriented programming language which supports network computing.

译文：____。

翻译方法：______________________。

6．Sun found that Java could be used to create dynamic homepage, which gave new life to this project.

译文：____。

翻译方法：______________________。

7．Pressing the keys tells the computer what to do or what to write.

译文：____。

翻译方法：______________________。

8．Other types of machines that are often not considered as PCs are Macs and servers running UNIX or Linux.

译文：____。

翻译方法：______________________。

科技英语中的句子分析

科技英语中有大量长句，这些长句中往往含有若干分句和许多短语及其他修饰限定成分，长难句是读懂文章和翻译的主要障碍。要想读懂此类长难句必须对长句进行深入细致的分析，先理清主干，再层层明确各成分之间的语法关系和语义逻辑关系。所以，掌握句子分析对于科技英语的阅读十分重要。

1．句子的分类

句子的分类方式有两种：

根据句子的用途来分，分为陈述句、疑问句、祈使句和感叹句。

根据句子的结构来分，分为简单句、列句、复合句和并列复合句。

（1）简单句

句中只含有一个主语和一个谓语的句子称为简单句。例如：

I study English every day.

我每天学英语。

I bought a bike last year.

我去年买了一辆自行车。

（2）并列句

由连接词或“;”把两个或两个以上的简单句连在一起的句子称为并列句。常用的连接词有：and, but, so, or, or else, either ... or, neither...nor, not only...but also, yet, for，等等。例如：

She did the work and she did it well.

她做了这件工作而且做得很好。

The children can go with us, or they can stay at home.

孩子们可以和我们一起去，或者他们也可以待在家里。

Either she leaves, or I will!

要么她走，要么我走！

I have been to Beijing many times but my parents have never been there.

我去过北京多次，但我父母亲从没去过。

I heard a noise so I got out of bed and turned the light on.

我听到一声响，所以起床把灯打开。

Go quick, or else you'll miss the bus.

快走，否则你要赶不上公共汽车了。

The first one was not good, neither was the second (one).

第一个不好，第二个也不好。

Not only is he himself interested in the subject but all his students are beginning to show an interest in it.

不仅他自己对这门课程感兴趣，而且他所有的学生也开始显示出对这门课的兴趣。

He worked hard, yet he failed.

他工作努力，可是他失败了。

We can't go for it is raining.

我们不能走，因为正在下雨。

（3）复合句

句子中含有从句的句子称为复合句。例如：

Memory is sometimes referred to as temporary storage because it will be lost if the electrical power to the computer is cut off.

内存有时被称为临时存储器，因为如果计算机断电，它的存储内容就会丢失。

（4）并列复合句

含有复合句的并列句，称为并列复合句。例如：

Thomas wants to go to the party, but his wife says she is too tired.

托马斯想去参加舞会，但是他妻子说她很累。

2. 基本句型

英语的基本句型分为五种：

（1）主语＋系动词＋表语（S+V+P）

（2）主语＋谓语（S+V）

（3）主语＋谓语＋宾语（S+V+O）

（4）主语＋谓语＋间宾＋直宾（S+V+O+O）

（5）主语＋谓语＋宾语＋宾补（S+V+O+C）

S—Subject，主语；V—Verbal phrase，谓语；P—Predicate，表语；C—Complement ，补语；O—Object，宾语。

现以简单句为例来举出五种基本句型的例子：

（1）主语＋系动词＋表语（S+V+P）

This　is　an English-Chinese dictionary.

The dinner　smells　good.

The football　is　on the floor.

（2）主语＋谓语（不及物动词）(S+V）

He has gone.

（3）主语＋谓语（及物动词）＋宾语（S+V+O）

They　knew　the answer?

He　enjoys　reading.

He　has refused　to help them.

（4）主语＋谓语（及物动词）＋间宾（多指人）＋直宾（多指物）(S+V+O+O）

My parent　bought　me　a gift.

He　showed　me　how to run the machine.

（5）主语＋谓语＋宾语＋宾补（S+V+O+C）

The song　made　him　a star.

They　found　the house　deserted.

He　asked　me　to come back soon.

I　saw　her　watching TV.

以上例句中的主语、宾语、表语都是由单词或词组来承担的，也可以由从句来承担，就变成了复合句。如果把这些句子再用并列连词连接起来，就变成了并列句或并列复合句。

3．句子成分

句子成分可分为主语、谓语、宾语、表语、定语、状语、补语、同位语。它们可以由单词来担任，也可以由词组或句子来担任。

（1）主语

可以作主语的有名词（如 boy）、主格代词（如 you）、数词、动词不定式、动名词、主语从句等。主语一般放在句首。例如：

名词作主语：The boy comes from America.

代词作主语：He made a speech.

数词作主语：Two and two is four.

主语从句作主语：What I know is important.

动名词作主语：Smoking is bad to health.

动名词短语作主语：Arguing with him is a waste of time.

不定式作主语：To pass the exam needs hard work.

主语从句作主语：That the moon goes around the earth is true.

对于后三个例子，动名词短语、动词不定式和 that 引导的主语从句作主语的情况，经常用 It 形式主语替代，写成如下形式：

It is a waste of time arguing with him.

It needs hard work to pass the exam.

It is true that the moon goes around the earth.

注 意

that 引导的主语从句作主语时，一般都用 It 形式主语来替代。What 引导的主语从句不用形式主语。而其他的像由 whether, when, where, who, how 等引导的主语从句可以用 It 形式主语，也可以不用。动词不定式和动名词短语作主语常用 It 形式主语替代，但也可不用。

（2）谓语

谓语是用来说明主语做了什么动作或处在什么状态。谓语由动词来担任，是英语时态、语态变化的主角，一般在主语之后。例如：

She speaks English fluently.

The child has been brought up by his mother.

（3）宾语

宾语是谓语动词动作的承受者，一般同主语构成一样，宾语可以由名词或起名词作用的成分担任，包括：名词、代词宾格、动名词、不定式、宾语从句。宾语一般放在谓语动词后面。例如：

名词作宾语：The girl needs a pencil.

Smoke made the city ugly.

代词作宾语：They did it yesterday.

I saw them getting on the bus.

名词、代词作双宾语：I showed him my pictures.

动名词作宾语：I like swimming.

不定式作宾语：I like to swim this afternoon.

宾语从句作宾语：We think (that) you are right.

（4）表语

表语是用来说明主语的性质、身份、特征和状态。表语一般放在系动词之后，和系动词一起构成句子的复合谓语。表语可以由名词、形容词、数词、介词短语、不定式、动名词、分词、表语从句等担任。例如：

名词作表语：This is an English-Chinese dictionary.

形容词作表语：These desks are yellow.

数词作表语：She is ten.

介词短语作表语：The book is in the bag.

不定式作表语：Her dream is to be a teacher.

动名词作表语：Your task is cleaning the windows.

现在分词作表语：The book is very interesting.

过去分词作表语：He was excited.

表语从句作表语：My question is how you knew him.

另外，系动词除了 be 动词之外，还有一些表“转变为”之意的动词和感官动词。例如：

He became a teacher at last.

他最终成了一名教师。

The egg smells bad.

这个鸡蛋难闻。

（5）定语

定语是对名词或代词起修饰、限定作用的词、短语或句子。充当定语的有形容词、名词、代词、数词、副词、介词短语、不定式短语、分词和定语从句等。有些定语要放在所修饰词的前面（前置），有些则要放在所修饰词的后面（后置）。需要后置的有形容词短语、介词短语、分词短语、不定式短语、定语从句、副词、不定代词（如 something 等）。例如：

形容词作定语：That is a beautiful flower.

形容词短语作定语：This is a subject worthy of careful study.（后置）

名词作定语：It is a ball pen.

代词作定语：This is my book, not your book.

数词作定语：There are two boys in the room

副词作定语：The boy there told me how to get to the station.（后置）

介词短语作定语：The girl in the classroom is my sister.（后置）

不定式作定语：I don't know the way to get to the station. （后置）

分词作定语：Our country is a developing country.

分词短语作定语：The TV set made in that factory is very good. （后置）

定语从句作定语：The person who is standing under the tree is my teacher. （后置）

（6）状语

状语用来修饰动词、形容词、副词或全句，说明事物发生的时间、地点、原因、目的、结果、方式、条件、伴随情况、程度等情况。状语可以由副词、介词短语、分词短语、不定式以及状语从句来担任，在句子中的位置很灵活。例如：

副词作状语：I often read the newspaper at night.

He speaks English very well.

This room is much bigger than that one.

介词短语作状语：On Sundays, there is no student in the classroom.

不定式作状语：He is studying hard to pass the exam.

状语从句作状语：While John was watching TV, his wife was cooking.（时间状语从句）

Wherever you go, you should work hard.（地点状语从句）

I'm late because I missed the bus.（原因状语从句）

The boss asked the secretary to hurry up with the letters so that he could sign her.（目的状语从句）

He got up so early that he caught the first bus.（结果状语从句）

You will certainly succeed so long as you keep on trying.（条件状语从句）

The old man always enjoys swimming even though the weather is rough.（让步状语从句）

She behaved as if she were the boss.（方式状语从句）

She is as bad-tempered as her mother.（比较状语从句）

分词短语作状语：He went out of the classroom, talking and laughing.（表伴随）

Hearing the news， they all jumped with joy.（表时间）

Not knowing how to work out the difficult physics problem, he asked the teacher for help.（表原因）

Given more time, they would have done it better.（表条件）

Helped by their teacher, the students finished the task successfully.（表方式）

He fired, killing two flying birds. （表结果）

Gaining much money, he still felt unhappy.（表让步）

分词短语作状语注意以下几点：

① 分词短语作状语可以看作是从各类状语从句转换而来的。如果状语从句中的谓语动词为被动结构，就用过去分词；如果状语从句中的谓语动词为主动结构，就用现在分词。同样可表示：原因、结果、条件、让步、时间、方式和伴随情况。

② 有时为了强调，分词前可带连词 when, while, if, though, as if, unless 等一起作状语，以便使句子的意思更清楚、更连贯。例如：

When doing his homework, the girl was listening to her classical music.（时间状语）

If given more time and money, we could have completed the task.（条件状语）

Although gaining much money, he still felt unhappy.（让步状语）

③ 分词的逻辑主语必须与主句的主语一致。这一点是最根本的原则。只有当两者一致时，分词作状语才能成立。如果不一致，就不能使用分词作状语，要在分词前面加上其逻辑主语，变成“独立主格结构”予以代替。例如：

Using negotiation instead of arm force, the two nations eventually solved the border dispute peacefully.（分词短语作状语）

(With) the peaceful means used, the two nations eventually solved the border dispute.（独立主格结构作状语）分词 used 与它前面的 means 有逻辑主谓关系。

分词独立主格结构作状语：它不是一个完整的句子，但却表达了一个完整的意义，有以下几种形式：

a. 名词/主格代词+分词

例如：Weather permitting (= If weather permits), we will go for a picnic tomorrow.

Her glasses broken (= Because her glasses were broken), she couldn't see the words on the blackboard.

b. 连接词+名词/主格代词+分词

例如：If the weather permitting, we would go outside for a picnic.

After the work done, we will have a relatively long vacation.

c. with+名词/主格代词+分词，该结构一般只用来表示原因状语。

例如：With his homework having been done, he went out for playing basketball.

With the bridge to be completed, the communication between the two cities will surely be strengthened.

（7）补语

补语（宾语补足语）位于宾语之后，是对宾语作出说明的成分。宾语与其补足语有逻辑上的主谓关系，它们一起构成复合宾语。能够充当宾语补足语的有：名词、形容词、介词短语、不定式、分词、副词等。一般情况下，宾补通常紧跟在宾语之后。例如：

名词作宾补：The war made him a soldier.

形容词作宾补：I find learning English difficult.

介词短语作宾补：I often find him at work.

不定式作宾补：The teacher asked the students to close the windows.

分词作宾补：I saw a cat running across the road.

副词作宾补：I saw the kite up and down.

（8）同位语

同位语是在名词或代词之后并列名词、代词或同位语从句，以此对前者加以说明，近乎于后置定语。例如：

We students should study hard.（students 是 we 的同位语，都是指同一批“学生”）

We all are students.（all 是 we 的同位语，都指同样的“我们”）

The news that our women volleyball team had won the championship encouraged us all greatly.（that 引起的从句是 news 的同位语）

分析句子结构时注意掌握以下几个问题：

① 先找主句，层层分解。浏览全句，先把句子的主干找出来，看看是属于哪一种基本句型。然后再分解其他的成分。例如：

One of the principal advantages of object-oriented programming techniques over procedural

主语

programming techniques is that they enable programmers to create modules that do not need

系词 表语

to be changed when a new type of object is added.

主句：One of the principal advantages is that...是“主系表”句型（S+V+P）。

分析 that 表语从句内部：

that they enable programmers to create modules (that do not need to be changed when a

主语 谓语 宾语 宾补 定语

new type of object is added).

基本句型:主语+谓语+宾语+宾补（S+V+O+C）

再分析 modules 后的定语从句：

that do not need to be changed when a new type of object is added.

主语 谓语 宾语 状语

基本句型：主语+谓语+宾语（S+V+O）

再分析 when 从句内部：

when a new type of object is added

主语 谓语

基本句型：主语+谓语（S+V）

其他的 of object-oriented programming techniques 介词短语作定语；over procedural programming techniques 介词短语作状语。

② 特别注意“后置定语”。 一个名词后面若紧跟着一个从句、分词短语、形容词短语、介词短语等，很有可能是它的后置定语。尤其修饰主语中心词的后置定语应该在主语和谓语中间。仍看上例：

advantages (of object-oriented programming techniques)

名词 介词短语作后置定语

modules (that do not need to be changed when a new type of object is added.)

名词 定语从句作后置定语

③ 一个句子只能有一个谓语动词。其他动词只能以非谓语动词形式出现，即现在分词、过去分词、不定式。例如：

(In the middle of the room) there was a Christmas tree (decorated with colored lights

状语 谓语 主语 定语

and glass balls).

注 意

①主句是 there be 结构。此结构中 there 是引导词，不作任何成分，be 是谓语，be 后面的名词是主语。此结构是倒装结构。②浏览句子找谓语时，可见句中有两个词 was 和 decorated 有成为谓语动词的可能性，was 是 be 动词的过去式形式，只可能是谓语动词；decorated 可能是过去式也可能是过去分词；因为一个句子中只能有一个谓语动词，既然 was 只能是谓语动词，那 decorated 就只能是过去分词——这种非谓语动词形式，作定语或状语。

④ 注意句子判断。完整的一句话必须是以句号、分号结束的。若见到并列连词 and，but 等连接两个句子，前后句要分开分析其主干结构。同时注意，and 和 but 等还可以连接两个词或词组。要搞清楚，它连接的是两个句子还是两个词或词组。例如：

(As we have mentioned before), a program (written in assembly language) is input to the
状语　主语　定语　谓语　宾语

assembler,(which translates the assembly-language instructions into machine code) and the
定语　主语

machine code, (which is the output from the assembler), is then executed.
定语　谓语

主句：a program is input to the assembler **and** the machine code is then executed
主语　谓语　宾语　主语　谓语

基本句型：主语+谓语+宾语（S+V+O）**and** 主语+谓语（S+V）

练　　习

Ⅰ．标注下列句子中画线部分的成分

1．A program is a list of instructions or statements for directing the computer to perform a required data processing task.

2．In other words, the only programming instructions that a computer actually carries out are those written using machine language.

3．A USB flash drive consists of a memory chip encased in a small piece of plastic with a USB connector on the front.

4．This simplifies the programmer's task, resulting in more reliable and efficient programs.

5．Despite the use of simpler structures at the logical level, some complexity remains, because of the large size of the database.

6．For developers, Web server software is freely available that can respond to requests for both documents and programs.

7. Between the two lies software called middleware, usually developed with a Web server-side scripting language that can interact with the DBMS, and can decode and produce HTML used for presentation in the client Web browser.

8. Filling out a Web page form, selecting an option from a menu displayed in a Web page, or clicking an onscreen ad are common ways database requests are made.

9. It is contained on a single chip called the microprocessor and the microprocessor is often contained within a cartridge that plugs in to the motherboard.

10. This step-by-step operation is repeated over and over again at an awesome speed till the program is performed.

II. 分析下列句子结构

1. Since many users of database systems are not deeply familiar with computer data structures, database developers often hide complexity through the following levels to simplify users' interactions with the system.

2. The next higher level of abstraction describes what data are stored in database, and what relationships exist among those data.

3. The logical level of abstraction is used by database administrators, who must decide what information is to be kept in the database.

4. It can relate data stored in one table to data in another as long as the two tables share a common data element.

5. We will introduce the devices which enable people and computer to communicate.

6. If the computer needs to fetch data from memory again, it is necessary for it to visit memory once more.

7. A printer transfers what you see on the monitor onto paper, using impact or non-impact printing technology.

8. Memory is an area that holds programs processed presently and data used by programs.

9. A program written in a high-level language can run on any computer that has an appropriate compiler for the language.

10. Information in it concerns some essential instructions that are required whenever we turn on the computer.

科技英语中的后置定语

所谓“定语”，顾名思义，就是起修饰限定作用的句子成分。在英语中，定语是用来修饰、限定名词或代词的词、短语或句子；在汉语中常译为“……的”。充当定语的主要是形容词（adjective）；此外，名词（noun）、代词（pronoun）、数词（numeral）、动名词（gerund）、分词（participle）、动词不定式（短语）（infinitive）、介词短语（preposition phrase）和定语从句（attributive clause）也都可以作定语。根据定语与其修饰的名词或代词的位置关系，可以分为前置定语和后置定语。科技英语中常用的后置定语主要有定语从句、分词和介词短语。

1．定语从句

定语从句必须置于它所修饰的名词（或代词）之后，被修饰的名词（或代词）称为先行词。先行词和从句之间用关系词衔接，常用的有：关系代词——that，which，who，whom，whose 和关系副词——when，where，why，how。这些关系词除了起引导作用以外，还在从句中充当一定的句子成分，例如：Software refers to programs that control the operation of the hardware.（软件是指控制硬件操作的程序。）中定语从句的连接词 that 就在从句中作主语。除作主语外，关系词还可以在从句中作宾语、定语和状语。

根据从句与先行词的语义关系，定语从句又分为限制性定语从句和非限制性定语从句。两者的区别主要有以下几点：

（1）形式不同

限定性定语从句主句和从句之间不用逗号隔开，口语中使用时也不停顿；而非限定性定语从句与主句之间通常用逗号隔开。

（2）作用不同

限定性定语从句用于对先行词的意义进行修饰、限制和识别，如果去掉，就会造成句意不完整或概念不清；而非限定性定语从句用于对先行词起补充说明作用，如果省略，句意仍然清楚、完整。

（3）关系词不同

关系词 that 和 why 通常不用于非限制性定语从句；另外，限制性定语从句中关系词有时可以省略，而非限制性定语从句中关系词一律不省略。

（4）先行词不同

限定性定语从句的先行词只能是名词或代词，而非限定性定语从句的先行词则可以是名词或代词，也可以是短语或句子。

（5）翻译方法不同

一般把限定性定语从句翻译在它所修饰的先行词之前，而把非限定性定语从句与主句分开。例如：

① Here we will introduce the devices which enable people and computer to communicate.

以下将介绍能使人机交流的设备。

② Music can be input into a PC through MP3 players or MP4 players, which are also output devices.

音乐可以通过 MP3 或 MP4 输入个人计算机，MP3 或 MP4 同时又是输出设备。

③ The motherboard is located at the bottom of the desktop case or the side of a tower case, on which there are several sockets.

主板位于台式机箱的底部或塔式机箱的侧面，它上面有很多凹槽。

④ Except for the PC world, where Windows dominates, almost every hardware vendor offers UNIX as its primary or secondary operating system.

除了在Windows操作系统占垄断地位的个人计算机领域，几乎所有的硬件销售商都把UNIX作为首选或第二操作系统。

注 意

定语从句的详细翻译技巧见“科技英语的翻译技巧”。

2. 分词作后置定语

分词有现在分词（v.+ing）和过去分词（v.+ed）两种。现在分词表示主动或进行；过去分词表示被动或完成。

分词具有形容词的性质，在科技英语中，最常见的是分词作定语。现在分词修饰的是发出该动作的名词，即与名词是主谓关系；过去分词修饰的是承受该动作的名词，即与名词是动宾关系。例如：Please give this form to the man *sitting* there. 现在分词sitting修饰的是发出sit（坐）这一动作的名词man。再例如A girl *called* Jenny wants to see you. 过去分词called修饰的是承受call（取名为）这一动作的名词girl。

当单个分词作定语时，放在被修饰的名词之前，这种用法在科技英语中相对较少；分词短语作定语时，放在被修饰的名词之后，即作后置定语。若是现在分词短语作后置定语，通常将分词短语翻译为“……的”放在被修饰的名词前；若是过去分词短语作后置定语，通常将分词短语译为“由/被……的”放在被修饰的名词前。例如：

① It is contained on a single chip called the microprocessor…

它（CPU）装在一个被称作微处理器的芯片上……

② Cards are electronic parts made up of chips.

卡是由多个芯片构成的电子元件。

③ In many cases, the graphics technology used in today's best games will show up in tomorrow's business applications.

在许多情况下，当今最好的游戏软件中使用的图形技术将被用于未来的商业软件中。

④ To perform word processing, you need a computer, a special program called a word processor, and a printer.

为执行文字处理，需要一台计算机、一种被称作文字处理器的专用软件，还有一台打印机。

3. 介词短语作后置定语

为避免头重脚轻，介词短语作定语时通常置于被修饰的名词或代词后面，在汉语译文中将其译为“……的”放在被修饰名词或代词前面。例如：

① Generally, the traditional 101-key keyboard has four key groups: the function key row at the top of the keyboard…

一般来说，传统的101键的键盘有四个键区：位于键盘顶部的功能键区……

② A monitor is hardware with a television-like viewing screen.

显示器是指具有类似电视机的显像屏幕的硬件。

③ This section presents some important components of computer hardware.

本章节介绍计算机硬件的几个重要组成部分。

④ Output devices are instruments of interpretation and communication between humans and computer system.

输出设备是人与计算机系统之间理解和交流的工具。

练　　习

指出下列句子中的后置定语及其修饰词

1. A computer is a fast and accurate system that is organized to accept, store and process data, and produce results under the direction of a stored program.

后置定语及其修饰词：________________________________

____

2. It is contained on a single chip called the microprocessor and the microprocessor is often contained within a cartridge that plugs in to the motherboard.

后置定语及其修饰词：________________________________

____

3. We will introduce the devices which enable people and computer to communicate.

后置定语及其修饰词：________________________________

____

4. Memory is an area that holds programs processed presently and data used by programs.

后置定语及其修饰词：________________________________

____

5. Like the input devices, output devices are instruments of interpretation and communication between humans and computer systems.

后置定语及其修饰词：________________________________

____

6. This is a major reason why personal computers are so popular today.

后置定语及其修饰词：________________________________

____

科技英语中的被动语态

随着科学技术日新月异的发展，科技英语与普通英语已分别成为一门独立的学科。科技英语具有陈述客观准确、逻辑结构严密、专业术语性强、语气正式等语言特点，在科技文章中，多数句子是无人称被动句，因为科技著作者想要客观地对待事物，而不强调行为的主体，所以通常不用 I，you，the operator 等作为句子开头。因此，广泛运用被动语态是科技英语的一个突出语法特征。被动语态的使用范围非常广泛：当着重指出动作的承受者或不必说出主动者、不愿说出主动者、无从说出主动者或是为了便于连贯上下文等场合都可以使用被动语态。在翻译科技英语文章时，英语的被动语态一般都译成汉语的主动句式，只有在特别强调被动动作或特别突出被动者时才译成汉语被动句式或无主句。

被动语态由助动词be加及物动词的过去分词构成。

以下归纳总结了基本的被动语态翻译处理技巧。

① 仍译成汉语被动句。原文主语在译文中仍作主语，译文中往往加入“被”“由”等字词。

At the physical layer, the information is placed on the physical network medium and is sent across the medium to another system.

在物理层，信息被放置到物理介质上并穿过介质被发送到其他系统。

② 变被动形式为主动形式，译成汉语主动句。英语中的被动句译成汉语主动句有以下几种不同情况：

a．直接翻译成汉语主动句。

There is a limit on the number of computers that can be attached to a single LAN.

对连入单个局域网的计算机数目有限制。

Nothing before could be compared with the Internet as long as the revolutionary it brings to the computer and communication world is concerned.

因特网带给计算机和通信世界的革命，之前任何事物都无法与之比拟。

The line between a private and public network has always been drawn at the gateway.

专用网和公用网之间通过网关连接。

Worms are characterized by having the ability to replicate themselves.

蠕虫病毒的特点是有能力复制本身。

b．原文主语在译文中仍作主语，译文中往往加入“可……”“用于……”“是……的”等字词。

Internet technology can be used with virtually any computer platform.

实际上因特网技术可用于任何计算机平台。

Viruses are classified into two major categories: boot viruses and file viruses.

病毒可分为主要的两类：引导扇区病毒和文件病毒。

Most vendors provide patches that are supposed to fix bugs.

大多数厂商都提供补丁，用来修补存在于产品中的缺陷。

Switching is performed at layer 2 of the seven-layer model.

交换技术是在七层模型的第二层上实现的。

Most viruses today are transmitted through the Internet.

当今很多病毒是通过因特网来传播的。

③ 译成无主句，被动语态中的主语被翻译成宾语，使表述更为通顺、流畅。

Firewalls are configured to protect against unauthenticated interactive logins from the “outside” world.

为防止来自“外部”世界未经授权的登录，配置了防火墙。

To prevent unauthorized access, some type of identification procedure must be used.

为防止非授权访问，必须采取某些识别措施。

④ 若没有动作的执行者，翻译时添加“我们”“人们”等。

There are several programming languages that are known as AI languages.

有好几种编程语言都是我们知道的 AI 专用编程语言。

⑤ 科技英语被动句经常出现 by, from, in, for 等引导的作状语的介词短语，可把该介词短语翻译成动作的执行者，即句子主语，而原主语则翻译成宾语。

Software engineering applications are used in wide range of activities, from industry to entertainment.

从工业到娱乐很多领域都应用到软件工程。

Computers has been widely used in almost every field.

几乎所有领域都广泛使用计算机。

For each computer and service, password should be used.

每台计算机和每种服务都应该使用密码。

从上面这些句子的翻译中可以看出，科技英语被动句多数情况下译成汉语主动句，翻译时讲究忠实于原文，更注重“通顺”原则。为了使句子通顺流畅，不必拘泥于哪种方法。试比较下面句子的翻译：

Sent messages are stored in electronic mailboxes.

电子邮箱存储发送的信息。

发送的信息存储在电子邮箱里。

如果没有上下文的参照，上面两种翻译均可，都准确无误地表达了原文的意思。

练　习

把下列句子译成汉语，注意被动语态的译法

1．Honeypots, essentially decoy network-accessible resources, could be deployed in a network as surveillance and early-warning tools.

2．Such analysis could be used to further tighten security of the actual network being protected by the honeypot.

3．In bus topology, all devices are connected to a central cable, called the bus or backbone.

4．All packets can be correctly received at their destinations.

5．LANs can be connected via telephone lines and radio waves, to exchange scientific data, and to share computing power.

6．Users all over the world are connected to exchange data, news and opinions to share resources and communicate with each other.

7．All of the basic Internet tools, including Telnet, FTP, Gopher, and the World Wide Web—are based on the cooperation of a client and one or more servers.

8．If the requests from the Web browser are understood by the Web server, they should use the same standard protocol.

9．Every computer that communicates over the Internet is assigned an IP address.

科技英语中名词的各种语法功能

在科技英语文章中，为了使意思表达更明确、更简练，常常大量使用名词，这是科技英语不同于普通英语的一大特点。下面依次介绍名词充当主语、表语、宾语、补足语、定语、状语、同位语以及其他功能。

1．名词作主语

在陈述句中，名词作主语时，其位置一般均在谓语之前。

Hardware and software generally need to be upgraded over time.

硬件和软件在使用超过一定时间后，通常需要升级。

2．名词作表语

在陈述句中，名词作表语时，其位置一般都在连系动词之后。

The electron is a particle of negative electricity.

电子是带负电的粒子。

The kernel is the heart of the UNIX operating system.

内核是 UNIX 操作系统的心脏。

3．名词作宾语

在陈述句中，名词作宾语，其位置一般在及物动词或介词之后。

（1）及物动词的宾语

This multitasking capability enables users to be more productive.

这种多任务功能使用户效率更高。

Light produces a chemical change.

光产生化学变化。

（2）介词宾语

Software is another name for a program or programs.

程序也叫软件。

Under a certain condition, water turns into ice.

在一定条件下，水变成了冰。

4. 名词作补足语

① 作宾语补足语，其位置一般在宾语之后。

This characteristic makes sound cards effective parts.

这一特性使声卡成为很有效的部件。

We call CPU the brain of a computer.

我们把 CPU 称作计算机的大脑。

② 作主语补足语，位于谓语之后。

These laws are named Newton's laws of motion.

这些定律称为牛顿运动定律。

Steel is considered one of the most useful metals in industry.

钢被认为是工业中最有用的一种金属。

实际上，如果把名词作宾语补足语的句子改为被动句，宾语补足语也就成为主语补足语。请看下例：

We must make China a modern, powerful socialist country.（宾语补足语）

我们必须使中国成为一个社会主义现代化的强国。

China must be made a modern, powerful socialist country.（主语补足语）

中国必须建成为一个社会主义现代化的强国。

5. 名词作定语

在英语中，名词常用作定语，这种名词叫作定语名词。名词的所有格形式或通格形式都可以作定语。个别情况下，还可用复数形式作定语。

（1）名词所有格作定语，通常表示所属关系

Man's first invention was the wheel.

人类的第一个发现是轮子。

Today's operating systems are of much greater difference from those used before.

现在的操作系统与以前所使用的有很大的不同。

（2）名词通格作定语

power plant 发电厂　　system software 系统软件

device driver 设备驱动程序　　the iron and steel industry 钢铁工业

名词作定语从位置上可分为：

① 由单个名词或名词短语作前置定语。

Language translators are software that translates high-level languages into machine languages.

语言翻译器是将高级语言翻译成机器语言的软件。

We can make an order of magnitude improvement in gain.

我们可以把增益提高一个数量级。

② 有些情况下，一个名词短语还可放在某名词之后作后置定语（一般都是用来表示尺寸、大小等的名词短语）。这时，可看成是在该名词短语前省去了介词 of。

These factories produce integrated circuits the size of a finger-nail.

这些工厂生产只有手指头那么大小的集成电路。

They made hollow steel boxes the width of the bridge.

他们制作了宽度与桥宽相同的一些空心钢箱。

名词作定语时，一般用单数形式，但有时也会以复数形式出现。

communications satellite 通信卫星　　goods train 货车

chemicals company 化工公司　　parts list 零件目录

6. 名词作状语

名词或名词短语作状语，它可表示：

（1）时间

China launched a communications satellite last month.

中国上个月发射了一颗通信卫星。

（2）距离

Light travels 300,000 kilometers per second. 光每秒走三十万千米。

（3）长度

The resultant vector is 9.06 unites long.

合成矢量的长度为 9.06 个单位。

（4）重量

This machine weighs 300 pounds.

这台机器重 300 英磅.

（5）倍数或次数

The cord vibrated just sixty times in a minute.

绳子正好每分钟振动 60 次。

（6）度数

The calorie is the quantity of heat required to raise the temperature of a gram of water one centigrade degree.

卡路里是把一克水的温度提高一摄氏度所需的能量。

（7）程度

If a solid is heated a great deal, it will change to a liquid or even to a gas.

固体大量受热，就会变成液体，甚至变成气体。

需要特别注意的是：不带任何冠词的单个名词可作状语表示“方式”“方面”等，主要位于某些形容词或动词之前。这种状语实质上是由构词法演变来的。原来，在这种名词或动词之间有一个连字符形成一个组合词，但在现代科技英语中，往往把连字符省去了，变成了名词作状语的一种特殊形式。

All these tools have been heat treated.

所有这些工具都已经热处理过了。

The material in this volume has been class tested.

本书的内容经过了课堂使用检验。

7. 名词作同位语

名词可以作另一个名词、句子部分内容或整个句子的同位语，起进一步说明的作用，其位置一般在与之同位的另一名词或句子之后。

The word "work" has a special meaning in physics.

"功"这个词在物理学中有特殊物理意义。

The term mass refers to the amount of matter present in a body.

质量这一术语是指物体含有的物质的量。

① 与句子部分同位：在一个名词与句子一部分同位时，往往带有自己的定语短语或从句，并有逗号与句子隔开。它与之同位的往往是含有动词意义的名词短语或非谓语动词短语。

One of the principal applications of the diode is in the production of a DC voltage from an AC supply, a process called rectification.

二极管的主要用途之一，在于由交流电源产生一个直流电压，这一过程叫作整流。

② 与整个句子同位：名词与整个句子同位时，也往往带有自己的定语，并有逗号（或破折号）与整个句子隔开。

Most metals may be deformed considerably beyond their elastic limits, a property known as ductility.

大多数金属形变的程度可以远远超过它们的弹性限度，这一性质称为延性。

③ 名词作同位语时的一些特殊情况。

a. 名词作同位语时，可以由 or, namely, that is, such as, for example 等引入。例如：

Metals, such as silver and copper, are good conductors.

像银和铜这样的金属都是良导体。

This compound consists of three elements, namely, oxygen, hydrogen and nitrogen.

这个化合物有三种元素组成，即氧、氢和氮。

b. 少数情况下，由于主语的同位语比主语要长得多，或者为了强调这个同位语，一般把同位语位于主语之前，并用逗号与主语分开。这种同位语一般译成单独的一句。

A land of heavy forests, America also enjoys bountiful rivers and lakes.

美国是一个有着大量茂密森林的国家，她同时拥有大量的河流、湖泊。

练　　习

指出下列画线部分的语法功能

1．Often <u>desktop</u> computers and notebooks are part of a network.

2．A database does not present <u>information</u> directly to a user.

3．New concepts reported for the first time in the scientific literature are indexed at new chemical abstracts headings, <u>a policy identical with that for new chemical substances</u>.

4．Sugar is made of three elements, <u>carbon, hydrogen and oxygen</u>.

5．<u>Computers</u> understand the 0’s and 1’s of machine language.

6．The present process of making steel from iron is only about <u>100 years</u> old.

7．Their primary disadvantage is <u>size</u>.

8．People name this kind of programs <u>application software</u>.

参 考 答 案

科技英语的翻译技巧

1．译文：输入设备和输出设备有时候被统称为外围设备。
翻译方法：译成中文被动句。

2．译文：这三个部件——算数逻辑部件、控制部件和存储部件存在于任何大小的处理器中。
翻译方法：译成中文主动句。

3．译文：计算机是一个快速而精确的系统，它可以用来接收、存储和处理数据并在一个已存储的程序的指引下输出结果。
翻译方法：定语从句的分译法。

4．译文：这就是为什么如今个人计算机如此流行的一个主要原因。
翻译方法：定语从句的合译法。

5．译文：Java 语言是一个支持网络计算的面向对象程序设计语言。
翻译方法：定语从句的合译法。

6．译文：Sun 公司发现 Java 可以用来创造动态网络主页，这给该计划带来了生机。
翻译方法：定语从句的分译法。

7．译文：按键盘上的键能告诉计算机要做什么或写什么。
翻译方法：选择词义（key）和加词（键盘上的）

8．译文：其他不被称作个人计算机的机器包括苹果机和运行 UNIX 和 Linux 操作系统的服务器。
翻译方法：加词（操作系统）、减词（type,often）、词性转译（running）。

科技英语中的句子分析

Ⅰ. 标注下列句子中画线部分的成分

1. A program is a list of instructions or statements for directing the computer to perform a required data processing task.

定语

2. In other words, the only programming instructions that a computer actually carries

定语从句

out are those written using machine language.

定语

3. A USB flash drive consists of a memory chip encased in a small piece of plastic with a USB

定语

connecter on the front.

定语

4. This simplifies the programmer's task, resulting in more reliable and efficient programs.

状语（表结果）

5. Despite the use of simpler structures at the logical level, some complexity remains, because of

状语 主语 谓语

the large size of the database.

6. For developers, web server software is freely available that can respond to requests for both

定语从句（修饰 software）

documents and programs.

7. Between the two lies software called middleware, usually developed with a web server-side

主语 定语（修饰 software） 非限定性定语

sc ripting language that can interact with the DBMS, and can decode and produce HTML used for presentation in the client web browser.

定语

8. Filling out a Web page form, selecting an option from a menu displayed in a Web page, or clicking an onscreen ad are common ways database requests are made.

主语 定语从句

9. It is contained on a single chip called the microprocessor and the microprocessor is often

定语

contained within a cartridge that plugs in to the motherboard.

定语从句

10. This step-by-step operation is repeated over and over again at an awesome speed till the

状语 状语

program is performed.

状语从句

Ⅱ. 分析下列句子结构

1. Since many users of database systems are not deeply familiar with computer data structures,

原因状语从句

database developers often hide complexity through the following levels to simplify users' interactions

主语 状语 谓语 宾语 方式状语 不定式作目的状语

with the system.

2. The next higher level of abstraction describes what data are stored in database, and what

定语 主语 定语 谓语 宾语

relationships exist among those data.

3. The logical level of abstraction is used by database administrators, who must decide what

主语 定语 谓语 方式状语

information is to be kept in the database.

非限定性定语从句

定语从句内部：

who must decide what information is to be kept in the database

主语 谓语 宾语

4. It can relate data stored in one table to data in another as long as the two tables share a

主语 谓语 宾语 定语 宾补 定语 条件状语

common data element.

5. We will introduce the devices which enable people and computer to communicate.

主语 谓语 宾语 定语从句

which 引起的定语从句内部：

which enable people and computer to communicate.

主语 谓语 宾语 宾补

6. If the computer needs to fetch data from memory again, it is necessary for it to visit

条件状语从句作状语 系 表语 状语

memory once more. ——形式主语

it 形式主语所代的真正主语

7. A printer transfers what you see on the monitor onto paper, using impact or

主语 谓语 宾语 宾补

non-impact printing technology.

方式状语

8．Memory is an area that holds programs processed presently and data used by programs.

主语　系 表语　　　　　　　　　　　　　　　定语

that 定语从句内部：

that holds programs (processed presently) and data (used by programs)

主语　谓语　宾语　　　　定语　　　　宾语　　　　定语

9．A program written in a high-level language can run on any computer that has an appropriate compiler for the language.

主语　　　　定语　　　　谓语　　状语　　　　定语

10．Information in it concerns some essential instructions that are required whenever we turn on the computer.

主语　定语　谓语　　定语　　宾语　　　　　定语

that 引起的定语从句内部：

that are required whenever we turn on the computer.

主语　谓语　　　　　时间状语

科技英语中的后置定语

1．后置定语及其修饰词：定语从句 that is organized to accept, store and process data, and produce results under the direction of a stored program 修饰 system；介词短语 of a stored program 修饰 direction。

2．后置定语及其修饰词：过去分词短语 called the microprocessor 修饰 chip；定语从句 that plugs in to the motherboard 修饰 cartrige。

3．后置定语及其修饰词：定语从句 which enable people and computer to communicate 修饰 devices。

4．后置定语及其修饰词：定语从句 that holds programs processed presently and data used by programs 修饰 area；过去分词短语 processed presently 和 used by programs 分别修饰 programs 和 data。

5．后置定语及其修饰词：介词短语 of interpretation and communication between humans and computer systems 修饰 instruments。

6．后置定语及其修饰词：定语从句 why personal computers are so popular today 修饰 reason。

科技英语中的被动语态

1．蜜罐的主要作用是引诱网络访问资源，作为监测和早期预警工具在网络里使用。

2．这些分析可用来进一步加强被蜜罐保护的实际的网络安全。

3．在总线型拓扑结构中，所有设备都连接到一根中心电缆或中枢上。

4．这些信息单元在目的端被正确地接收。

5．各个局域网可通过电话线与无线电波连接，来进行科学数据的交换以及计算处理能力的共享。

6．世界各地的用户连入因特网来交换数据、新闻和观点，以达到资源共享、互相沟通。

7．所有基本的互联网工具，包括远程登录、FTP、Gopher、万维网，都建立在一个客户端和一个或多个服务器协同工作的基础之上。

8．Web 服务器要明白 Web 浏览器发出的请求，双方必须采用同一标准协议。

9．在互联网上进行通信的每台计算机都分配有一个 IP 地址。

科技英语中名词的各种语法功能

1．定语
2．宾语
3．同位语
4．同位语
5．主语
6．状语
7．表语
8．宾语补足语

参 考 文 献

[1] 陈枫艳. 计算机专业英语[M]. 北京：科学出版社，2006.

[2] TIMOTHY J O LEARY, LINDA I O LEARY. 计算机专业英语：Computing Essentials[M]. 2003 影印版. 北京：高等教育出版社，2003.

[3] DALE N, LEWIS J. 计算机科学概论：Computer Science Illuminated [M]. 英文版.3 版. 北京：机械工业出版社，2008.

[4] 余芳. 计算机专业英语[M]. 北京：冶金工业出版社，2006.

[5] 宋德富，司爱侠. 计算机专业英语教程[M]. 北京：高等教育出版社，2004.

[6] 王国超，王玉律，任煜昌. 计算机专业英语[M]. 北京：冶金工业出版社，2005.

[7] 刘兆毓. 边用边学计算机英语[M]. 北京：清华大学出版社，2007.

[8] 刘兆毓. 计算机英语[M]. 3 版. 北京：清华大学出版社，2003.

[9] 卜艳萍，周伟. 计算机专业英语[M]. 2 版. 北京：清华大学出版社，2008.

[10] 王志强. 计算机导论[M]. 北京：电子工业出版社，2007.

[11] 吴冰. 计算机英语[M]. 北京：航空工业出版社，2007.

[12] 赵克林，朱龙. 计算机英语[M]. 重庆：西南师范大学出版社，2006.

[13] 龚沛曾，陆慰民，杨志强. Visual Basic 程序设计简明教程[M]. 2 版. 北京：高等教育出版社，2004.

[14] 陈建铎. 计算机应用基础教程：For Windows 2000[M]. 3 版. 陕西：西安电子科技大学出版社，2006.

[15] 徐士良. 计算机软件技术基础[M]. 北京：清华大学出版社，2006.

[16] 刘艺，王春生. 计算机英语[M]. 北京：机械工业出版社，2005.

[17] 姜同强. 计算机英语[M]. 北京：清华大学出版社，2005.